Springer Series in **Materials Science** 12

Edited by Ulrich Gonser

T. Suzuki   S. Takeuchi   H. Yoshinaga

Dislocation Dynamics and Plasticity

With 166 Figures

Springer-Verlag

Berlin Heidelberg New York London Paris
Tokyo Hong Kong Barcelona Budapest

Professor Taira Suzuki
Department of Materials Science and Technology, Faculty of Industrial Science and Technology, Science University of Tokyo, Noda, Chiba 278, Japan

Professor Shin Takeuchi
Institute for Solid State Physics, University of Tokyo, Roppongi, Minato-ku, Tokyo 106, Japan

Professor Hideo Yoshinaga
Graduate School of Engineering Sciences, Kyushu University, Kasuga-shi, Fukuoka 816, Japan

Series Editors:

Prof. Dr. h. c. mult. *K. A. Müller*
IBM, Zürich Research Lab.
CH-8803 Rüschlikon, Switzerland

Prof. Dr. *U. Gonser*
Fachbereich 12/1
Werkstoffwissenschaften
Universität des Saarlandes
W-6600 Saarbrücken, Fed. Rep. of Germany

***M. B. Panish*, Ph. D.**
AT&T Bell Laboratories,
600 Mountain Avenue,
Murray Hill, NJ 07974, USA

***A. Mooradian*, Ph. D.**
Leader of the Quantum Electronics Group, MIT,
Lincoln Laboratory, P. O. Box 73,
Lexington, MA 02173, USA

Prof. *H. Sakaki*
Institute of Industrial Science,
University of Tokyo,
7-22-1 Roppongi Minato-ku,
Tokyo 106, Japan

Managing Editor: Dr. Helmut K. V. Lotsch
Springer-Verlag, Tiergartenstrasse 17
W-6900 Heidelberg, Fed. Rep. of Germany

Title of the original Japanese edition: Ten'ino Dainamikkusuto Sosei
© Shokabo Publishing Co., Ltd., Tokyo 1985

ISBN-13:978-3-642-75776-1   e-ISBN-13:978-3-642-75774-7
DOI: 10.1007/978-3-642-75774-7

Library of Congress Cataloging-in-Publication Data. Suzuki, Taira, 1918– Dislocation Dynamics and plasticity : with 166 figures / T. Suzuki, S. Takeuchi, H. Yoshinaga. p. cm.–(Springer series in materials science ; v. 12) Translation of: Ten'ino dainamikkusuto sosei. Includes bibliographical references and index. ISBN-13:978-3-642-75776-1(U.S.)1.Deformations (Mechanics)2.Plasticity.3.Dislocations in metals.4.Dislocations in crystals. I. Takeuchi, Shin, 1935– . II. Yoshinaga, Hideo, 1931– . III. Title. TA417.6.S8913 1990   620.1'123–dc20   90-10278

Preface

In the 1950s the direct observation of dislocations became possible, stimulating the interest of many research workers in the dynamics of dislocations. This led to major contributions to the understanding of the plasticity of various crystalline materials. During this time the study of metals and alloys of fcc and hcp structures developed remarkably. In particular, the discovery of the so-called inertial effect caused by the electron and phonon frictional forces greatly influenced the quantitative understanding of the strength of these metallic materials. Statistical studies of dislocations moving through random arrays of point obstacles played an important role in the above advances. These topics are described in Chaps. 2–4.

Metals and alloys with bcc structure have large Peierls forces compared to those with fcc structure. The reasons for the delay in studying substances with bcc structure were mostly difficulties connected with the purification techniques and with microscopic studies of the dislocation core. In the 1970s, these difficulties were largely overcome by developments in experimental techniques and computer physics. Studies of dislocations in ionic and covalent bonding materials with large Peierls forces provided information about the core structures of dislocations and their electronic interactions with charged particles. These are the main subjects in Chaps. 5–7.

In order to focus on the fundamentals of plasticity of crystalline materials, Chaps. 2–7 concentrate on a discussion of the yield strength of crystals at low temperatures. As a result, descriptions of many complex phenomena associated with interactions between dislocations, grain boundaries, multi-phase systems, etc., are omitted. From a practical viewpoint, these are unavoidable subjects; they are described in Chaps. 8–10 with respect to the deformation of materials at high temperatures. The materials treated in these chapters are metallic materials, including intermetallic compounds and metal–ceramic composites.

The present volume is a revision of the original Japanese edition, which was published by Shokabo in 1985. T. Suzuki, S. Takeuchi and H. Yoshinaga wrote Chaps. 2–4, 5–7 and 8–10, respectively. The introductory chapter was written by S. Takeuchi.

September 1989

Taira Suzuki
Shin Takeuchi
Hideo Yoshinaga

Contents

1. Dislocations and Their Fundamental Properties

In 1934, three researchers, Orowan, Taylor and Polanyi, independently published epochal papers in which the concept of crystal dislocation was clearly demonstrated. Thus, the year 1934 is generally recognized as the birth-year of the dislocation. During the following few years, the elasticity theory of dislocations was established. Since the 1950s, various new experimental techniques, such as electron microscopy, have been developed and have contributed to the understanding of dislocations in a variety of crystals. In this preparatory chapter, we briefly survey the concept of the dislocation and its fundamental properties. An introduction to this subject can be found in [1.1–3] and more advanced treatments in [1.4–7].

1.1 Geometry of a Dislocation

We can introduce a dislocation into a perfect crystal lattice in the following way: make a cut in the crystal along a lattice plane, $ABCD$ in Fig. 1.1, and then translate the upper part of the crystal with respect to the lower part by an atomic distance in a direction parallel to the cut, as shown by the arrow **b** in the figure, so that the two faces at the cut can be connected coherently again.

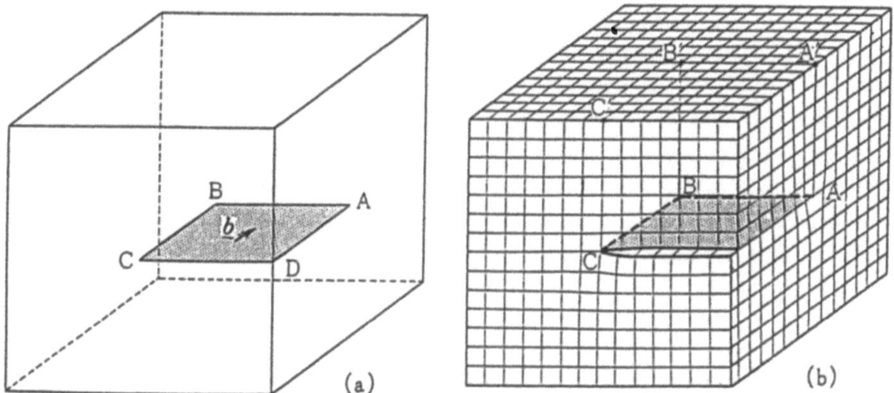

Fig. 1.1a,b. The process of introducing a dislocation into a crystal. $\overline{AB}$ in (b) is the edge dislocation and $\overline{BC}$ the screw dislocation

By this procedure, the crystal can be restored to the perfect lattice again except near the lines AB and BC; near these lines the lattice becomes wrinkled, as illustrated by lattice lines in Fig. 1.1b. These highly distorted lines in the crystal are called dislocation lines or simply dislocations. The dislocation AB is named an *edge dislocation*, after the edge-like shape of the lattice at the center of the dislocation. This type of dislocation can also be produced by inserting an extra lattice plane $AA'B'B$ in an otherwise perfect lattice. The inserted plane is called the extra half-plane. The dislocation BC is called a *screw dislocation*, because when we go around the dislocation line following a lattice plane perpendicular to the line, after every turn we come to the next lattice plane, like a screw motion. If we initially make a triangular cut ACD in Fig. 1.1a, we can introduce a dislocation CD, which is parallel to neither AB nor BC. Such a dislocation is called the mixed dislocation. The translation vector (**b** in Fig. 1.1a) is named the *Burgers vector*. For the purpose of representing a dislocation in an illustration, we customarily use an inverted T after the lattice configuration at the center of the edge dislocation.

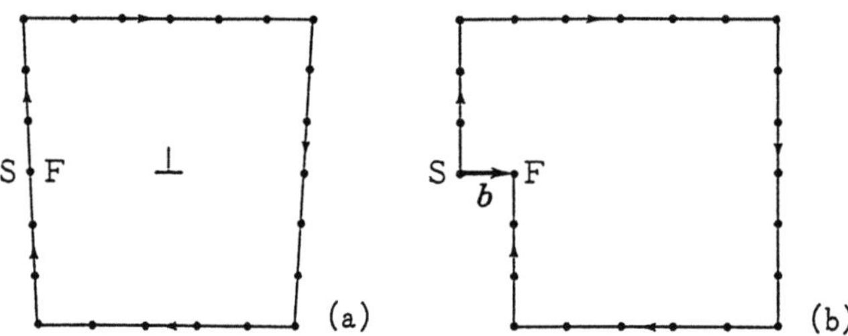

Fig. 1.2a,b. Definition of the Burgers vector b by use of a Burgers circuit

The Burgers vector of any dislocation can be defined, including the sense of it, as follows. Defining the direction of the dislocation line, we consider a closed circuit around the dislocation line in the direction of a right-hand screw, as shown in Fig. 1.2a in which the direction of the dislocation line is into the plane of the paper. Then we transfer the same circuit to a perfect lattice without the dislocation and as a result we have an offset between the starting point of the circuit S and the final point F, as shown in Fig. 1.2b. The offset vector $b = \overrightarrow{SF}$ is the Burgers vector and the circuit is the Burgers circuit. According to this definition of the Burgers vector, we can easily verify the following relation between the Burgers vectors b_i when more than three dislocations merge at a point (called the dislocation node):

$$\sum_i b_i = 0 , \tag{1.1}$$

assuming the directions of all the dislocations to be towards the node, see Fig. 1.3. The relation (1.1), which corresponds to the Kirchhoff rule in electric circuit theory, is called the conservation law of the Burgers vector. As a simple application

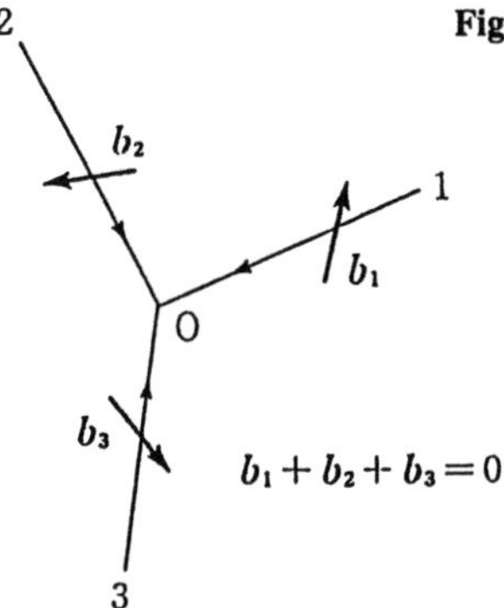

of this law, we can deduce the well-known fact that a dislocation cannot have a terminal point inside the crystal.

As may be imagined from the procedure of producing a dislocation mentioned above, a dislocation can be translated along the lattice plane, $ABCD$ in Fig. 1.1. This translational motion is called glide motion. The plane of the glide is confined to a plane containing both the dislocation line and the Burgers vector, and hence for an edge dislocation the glide plane is uniquely determined. For a screw dislocation, however, several lattice planes, in principle any plane containing the dislocation line, can be the glide plane; for example, the plane $BB'C'C$ in Fig. 1.1 can also be a glide plane. Such mutually intersecting glide planes for a screw dislocation are called the cross-slip planes. In order to translate an edge dislocation in the direction perpendicular to the glide plane, an addition or extraction of atoms is required along the dislocation line, and thus, such a motion, called the climb motion, cannot take place without atomic diffusion (absorption or emission of vacancies or interstitials). At high enough temperatures where the thermal diffusion of atoms becomes active, the climb motion of edge dislocations actually occurs in crystals.

The Burgers vector normally coincides with one of the lattice vectors of the crystal. In some crystals, however, dislocations with Burgers vectors smaller than the lattice vector can also exist. These dislocations are called partial dislocations. Since a stacking fault is necessarily formed on either side of a partial dislocation, a partial dislocation is stable only when the stacking fault can exist stably. Conversely, at the terminal end within a crystal of any stacking fault there must necessarily be a partial dislocation.

In general, the dislocation line tends to orient in a low-index direction on the glide plane so as to lower the self-energy of the dislocation. The potential energy of the dislocation fluctuating with the period of the lattice spacing is called the Peierls potential. A transition region of a dislocation from one potential valley to the next is called a kink, and that from one glide plane to its neighboring plane is called a jog, see Fig. 1.4.

When there are two or more dislocations close to one another, they tend to take a characteristic low energy configuration. Two edge dislocations with

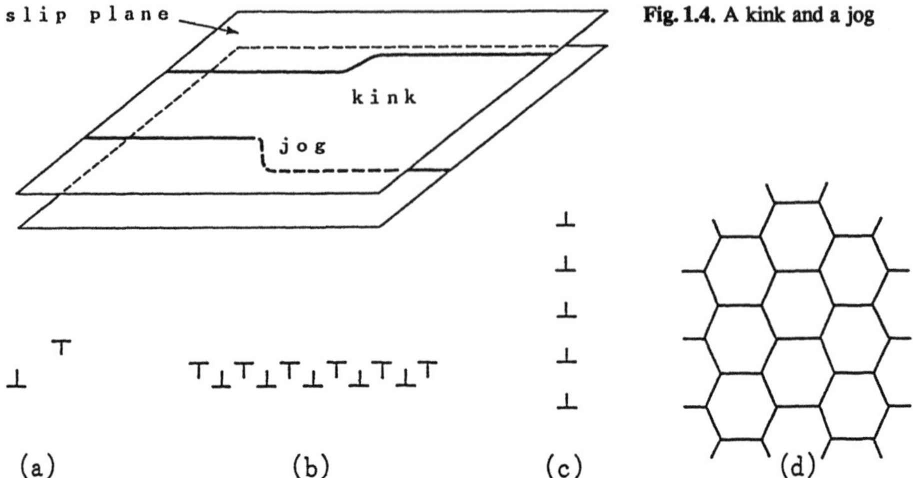

Fig. 1.5a–d. Various stable dislocation configurations: **(a)** dislocation dipole, **(b)** dislocation multipole, **(c)** polygon wall, **(d)** dislocation network

opposite signs take the stable configuration shown in Fig. 1.5a, which is called a dislocation dipole. When many dislocations are multiplied on parallel glide planes by plastic deformation, a dislocation multipole, shown in Fig. 1.5b is formed. Edge dislocations with an identical Burgers vector on different planes have a low energy when they arrange themselves vertically. Such an edge dislocation wall presented in Fig. 1.5c is often called a polygon wall. Dislocations with a different Burgers vector which pass through the glide plane of the dislocations being considered are named forest dislocations. These two kinds of dislocations sometimes react to form a third kind of dislocation with a Burgers vector $b_3 = b_1 + b_2$ at their intersection points. As a result of such a reaction between the two groups of dislocations, a configuraton with the network shape given in Fig. 1.5d is formed, called a dislocation network.

After a large plastic deformation of a crystal, a large number of dislocations are introduced into the crystal. These dislocations do not distribute uniformly, and many are localized in the form of a plate or a wall as a result of formation of multipoles and polygon walls, and thus a grain interior is divided into small regions by diffuse boundaries with a high density of dislocations. Such a structure is called the dislocation cell structure.

1.2 Stress Field and Energy of Dislocations

A characteristic feature of the dislocation is that it is accompanied by a long-range stress field and a large strain energy. At the very center of the dislocation, with a diameter of a few atomic distances, the lattice distortion is too large for elasticity theory to be applied. This region is called the dislocation core. The stress and strain fields outside the core region can be calculated by solving the

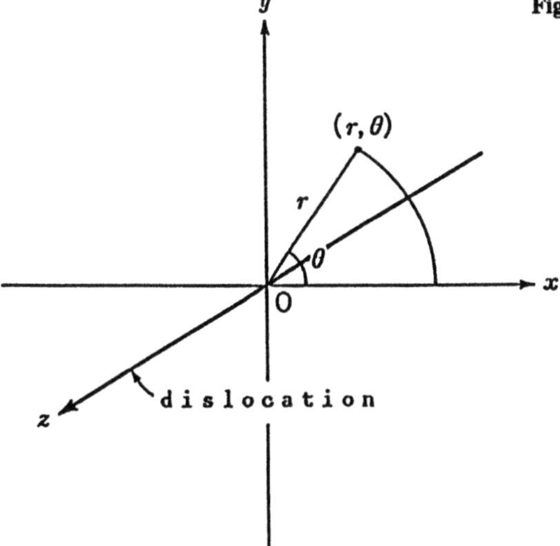

Fig. 1.6. Coordinate system for a dislocation

equilibrium equations for an elastic body. For simplicity, we assume that the crystal is elastically isotropic.

Take the z-axis along the dislocation line and the xz-plane on the glide plane (Fig. 1.6). For the screw dislocation, the total displacement of the lattice around the x-axis is equal to b (the magnitude of the Burgers vector), and hence the strain field can easily be obtained. Using cylindrical coordinates, we can write

$$\varepsilon_{\theta z} = \varepsilon_{z\theta} = \frac{b}{2\pi r} \ . \tag{1.2}$$

Let the shear modulus of the crystal be μ. The stress field is

$$\sigma_{\theta z} = \sigma_{z\theta} = \frac{\mu b}{2\pi r} \ , \tag{1.3}$$

$$\sigma_{r\theta} = \sigma_{zr} = \sigma_{rr} = \sigma_{\theta\theta} = \sigma_{zz} = 0 \ .$$

The hydrostatic pressure $p \equiv \frac{1}{3}(\sigma_{rr} + \sigma_{\theta\theta} + \sigma_{zz})$ is equal to zero. In Cartesian coordinates, the most important stress component is that of the shear stress along the slip plane or σ_{yz}, which is given by

$$\sigma_{yz} = \frac{\mu b}{2\pi} \frac{x}{x^2 + y^2} \ . \tag{1.4}$$

The stress field of the edge dislocation is not so easy to obtain as in the case of the screw dislocation. It has been shown that the following relations are the solutions, satisfying the equilibrium equations, for the edge dislocation with the Burgers vector parallel to the x-axis:

$$\sigma_{rr} = \sigma_{\theta\theta} = -\frac{\mu b}{2\pi(1-\nu)} \frac{\sin\theta}{r} \ ,$$

$$\sigma_{zz} = -\frac{\mu\nu b}{\pi(1-\nu)}\frac{\sin\theta}{r}\,,$$

$$\sigma_{r\theta} = \sigma_{\theta r} = \frac{\mu b}{2\pi(1-\nu)}\frac{\cos\theta}{r}\,,$$

$$\sigma_{rz} = \sigma_{zr} = 0\,,$$

$$p = \frac{1}{3}(\sigma_{rr} + \sigma_{\theta\theta} + \sigma_{zz}) = -\frac{\mu b}{3\pi}\left(\frac{1+\nu}{1-\nu}\right)\frac{\sin\theta}{r}\,. \tag{1.5}$$

Here ν is the Poisson ratio. A distinctive feature of the stress field of the edge dislocation is that it is accompanied not only by the shear stress components but also by a hydrostatic pressure, as may readily be understood from the atomic configuration. In Cartesian coordinates, the shear stress component parallel to the slip plane σ_{yx} is

$$\sigma_{yx} = \frac{\mu b}{2\pi(1-\nu)}\frac{x(x^2 - y^2)}{(x^2 + y^2)^2}\,. \tag{1.6}$$

The stress field of a mixed dislocation can be obtained by summing the stress field of the edge component and that of the screw component.

The total elastic strain energy stored around a dislocation can, in principle, be obtained by integrating local strain energies in elemental volumes. A simpler method of obtaining the strain energy is to calculate the work done in the process of creating the dislocation in the perfect lattice. Let us consider the process of introducing a screw dislocation in Fig. 1.6 by making a cut in the xz-plane ($x > 0$) and displacing the upper part by b in the z-direction. Taking account of the fact that the average stress during this procedure at a distance r from the dislocation is a half of the value $\sigma_{\theta z}$ in (1.3), the elastic energy stored in a cylindrical region outside the dislocation core with a radius $r_{\rm c}$ up to the outer cutoff radius R, per unit length of the screw dislocation, is calculated as

$$E_{\rm el}^{\rm s} = \frac{1}{2}\int_{r_{\rm c}}^{R}\frac{\mu b^2}{2\pi r}\,dr = \frac{\mu b^2}{4\pi}\ln\frac{R}{r_{\rm c}}\,. \tag{1.7}$$

In the same way, the elastic energy per unit length of the edge dislocation is obtained by integrating the value $\sigma_{\theta r}$ ($\theta = 0$) in (1.5) as

$$E_{\rm el}^{\rm e} = \frac{1}{2}\int_{r_{\rm c}}^{R}\frac{\mu b^2}{2\pi(1-\nu)r}\,dr = \frac{\mu b^2}{4\pi(1-\nu)}\ln\frac{R}{r_{\rm c}}\,. \tag{1.8}$$

A special feature of these results is that because both (1.7) and (1.8) diverge for $R \to \infty$, the elastic energy of a dislocation in an infinite crystal becomes infinite. The total energy of a dislocation is the sum of the above elastic energy calculated from elasticity theory and the energy associated with the dislocation core. The core energy is formally incorporated into the elastic energy by using a fictitious inner cutoff radius r_0 instead of $r_{\rm c}$. Thus, the total energy of the dislocation per unit length is generally expressed as

$$E = \frac{Kb^2}{4\pi} \ln \frac{R}{r_0} , \tag{1.9}$$

where K is a constant called the energy factor, which is equal to μ for the screw dislocation and $\mu/(1 - \nu)$ for the edge dislocation in an isotropic crystal. The value of r_0 is usually between $b/3$ and $b/4$.

In a real crystal, a number of dislocations arrange themselves so as to cancel the strain fields with each other. Consequently, the range over which the elastic field of each dislocation extends is about the spacing between dislocations. The dislocation density in well-annealed crystals is in the range 10^4–10^6 cm^{-2} and hence the value of R in (1.9) is $10^4 b$–$10^5 b$. Thus, the self-energy of a dislocation per unit length is generally expressed approximately by

$$E \sim \mu b^2 . \tag{1.10}$$

The self-energy per atomic distance of a dislocation is roughly μb^3, which amounts to a few eV to 10 eV. On the other hand the free energy decrease resulting from the entropy increase (configurational entropy plus vibrational entropy) due to the existence of a dislocation is at most 1 eV, even at the melting point. Consequently, a dislocation cannot exist under thermal equilibrium, unlike point defects.

From the relation that the energy of a dislocation is proportional to b^2, one can deduce the important fact that only a few types of dislocations can exist stably in any crystal. If a dislocation with a Burgers vector b_1 can be decomposed into two dislocations with Burgers vectors b_2 and b_3 ($b_1 = b_2 + b_3$) satisfying the relation $b_1^2 > b_2^2 + b_3^2$, then the b_1 dislocation is not stable for energy reasons. A simple example is that a dislocation with a Burgers vector of an integral multiple of a lattice vector should spontaneously decompose into unit dislocations with a unit lattice vector. As another example, let us consider possible dislocation types in the bcc lattice. The minimum Burgers vector is $\frac{1}{2}\langle 111 \rangle$ in units of the lattice constant. The Burgers vector $b_1 = [100]$ can be geometrically decomposed into $b_2 = \frac{1}{2}[111]$ and $b_3 = \frac{1}{2}[1\bar{1}\bar{1}]$, but due to the relation $b_1^2 < b_2^2 + b_3^2$ the dislocation with $b = \langle 100 \rangle$ generally exists in the crystal in a stable manner. The dislocation with $b_1 = [110]$ can generally reduce the energy by decomposing into $b_2 = \frac{1}{2}[111]$ and $b_3 = \frac{1}{2}[11\bar{1}]$ because $b_1^2 > b_2^2 + b_3^2$ in this case, and hence the dislocation with the Burgers vector $\langle 110 \rangle$ cannot generally exist in bcc crystals. In the above argument we have always added the adverb "generally", because the b^2 rule for the dislocation energy does not always hold, due to elastic anisotropy. Indeed, there are exceptions to the above rules for bcc metals.

A dislocation with a minimum Burgers vector or the minimum lattice vector can in some cases reduce its energy by splitting into partial dislocations. The splitting happens when the increase in energy due to the appearance of the stacking fault between the partials is smaller than the gain of the strain energy of the dislocation by the splitting. Such a dislocation is called an extended dislocation or a dissociated dislocation.

1.3 Force on a Dislocation

Due to the existence of a long-range strain field, the motion of a dislocation induces a change in the crystal shape. As a result, when a stress acts on a crystal, a force is exerted on the dislocation. Consider a case in which a shear stress τ is acting on a crystal containing an edge dislocation (Fig. 1.7a). The force f on the dislocation can be deduced by the amount of work done by the applied stress when the dislocation glides a certain distance. Assuming for simplicity that the sides of the crystal have a unit length, the shear strain brought about by the dislocation gliding a distance δ is given by $b\delta$, because the shear strain for the dislocation glide from left to right across the crystal is simply b. Consequently, from the definition of the force, $\tau b\delta = f\delta$. It follows that

$$f = \tau b \, . \tag{1.11}$$

The same force equation holds for the screw dislocation. The stress τ in (1.11) is assumed to be the shear stress component in the direction of the Burgers vector on the glide plane. It should be noted here that, since the force f on a dislocation always acts in the direction perpendicular to the dislocation line, the direction of the shear stress and the force on the dislocation are perpendicular to each other for the screw dislocation.

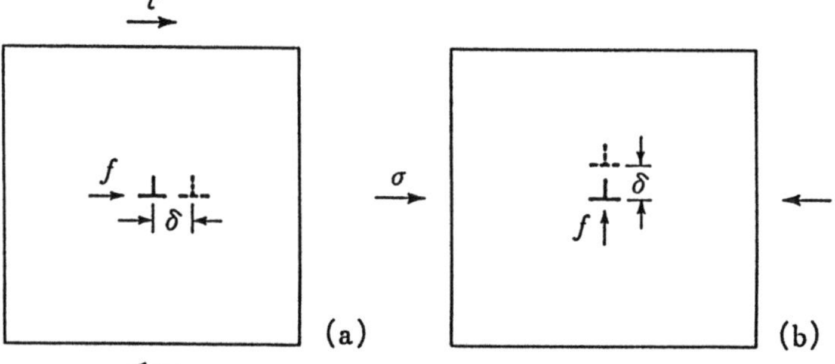

Fig. 1.7a,b. The force on a dislocation due to applied stress

When a compressive or tensile stress is acting on side faces as shown in Fig. 1.7b, a climb force is exerted on the edge dislocation. The climb force, derived similarly to the glide force, is

$$f = \sigma b \, . \tag{1.12}$$

A general expression for the force exerted on a dislocation by an arbitrary stress can be derived in the following way. Consider that a dislocation segment dl moves a distance dx under the action of stress tensor P, as shown in Fig. 1.8. The elemental force acting on elemental area $ds = dl \times dx$ is dsP. Because the area ds moves by b with the migration of the dislocation, the work done dW

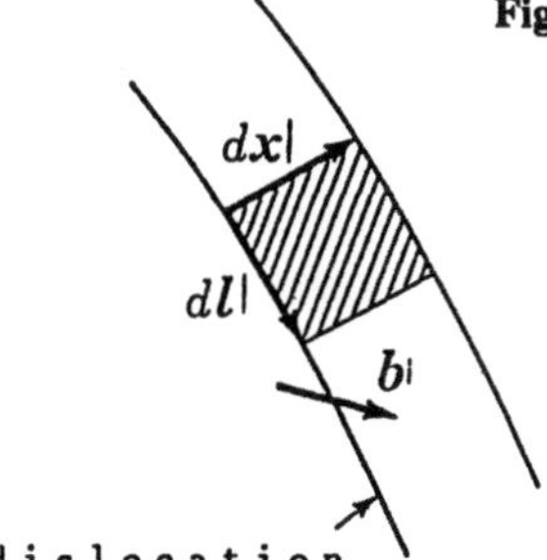

Fig. 1.8. Obtaining the Peach-Köhler force

during this migration is $dW = (ds\,P) \cdot b$. Using the fact that the tensor P is symmetric, we can transform the dW in the following way:

$$dW = (ds\,P) \cdot b = (bP) \cdot ds$$
$$= (bP) \cdot (dl \times dx)$$
$$= \{(bP) \times dl\} \cdot dx \ .$$

Thus, the force df on the dislocation element is obtained, by the definition $dW = df \cdot dx$, as

$$df = (bP) \times dl \ , \tag{1.13}$$

which is called the *Peach-Köhler equation.*

An internal stress also exerts a force on a dislocation. An example is the interaction between dislocations. The stability of the dislocation configuration shown in Figs. 1.5a and c can be derived from the stress field around the edge dislocation given by (1.4).

It is known that the dislocations in fcc and hcp metals are dissociated into partial dislocations. The width of these extended dislocations is determined by the balance between the repulsive force resulting from the interaction of two partials and the surface tension of the stacking fault (i.e. the stacking fault energy). Using this fact, we can estimate the stacking fault energy in various crystals rather accurately by measuring the width of extended dislocations by transmission electron microscopy.

1.4 String Model of a Dislocation

Because a dislocation is a flexible, one-dimensional entity, it is difficult to treat its motion strictly. Hence, the string model of dislocations finds widespread application. In this model the dislocation is treated as if it possesses a proper mass and a tension, in spite of the fact that the dislocation is nothing but a distorted state of the lattice.

Compared with the strain field around a dislocation at rest, that of a moving dislocation shrinks in the direction of its motion. This shrinkage is rather easily derived by the Lorentz transformation of the equation of motion for the elastic body. One can calculate the increase in the elastic energy of the moving dislocation from the shrinkage of the strain field. The effective line mass A of the dislocation is defined by the inertial mass corresponding to the above elastic energy increase. When the dislocation velocity v is much smaller than the velocity of the elastic shear wave c, the energy of the dislocation is expressed as

$$E = \frac{E_0}{\sqrt{1 - \frac{v^2}{c^2}}} \fallingdotseq E_0 \left(1 + \frac{v^2}{2c^2} \right) . \tag{1.14}$$

where E_0 is the rest energy. Equating the kinetic energy of the dislocation $Av^2/2$ to $E_0 v^2/2c^2$, one obtains

$$A = \frac{E_0}{c^2} . \tag{1.15}$$

Substituting the relations $E_0 \simeq \mu b^2$ and $c = \sqrt{\mu/\varrho}$,

$$A \simeq \varrho b^2 , \tag{1.16}$$

where ϱ is the density of the crystal. This result means that the effective line mass corresponds to a real mass of a filament of the material with a diameter of an atomc distance. This value is extremely small for the magnitude of the force usually exerted on a dislocation and hence the acceleration of the dislocation is instantaneous.

Suppose that a dislocation becomes undulated with a wavelength λ. Then it tends to straighten because the elastic energy of the dislocation is increased by the undulation. Approximating that the energy increment is proportional to the length increment, the dislocation is regarded as having a constant line tension C. Namely, for curvature κ a restoring force κC is exerted. Far away from the dislocation, i.e. at distances much greater than the wavelength λ, the elastic field is substantially unaffected by the undulation and hence the increase in energy for the increased length of the dislocation is less than in (1.9) and may be approximated by replacing R by λ in the equation. If λ is not too small, the line tension C can be roughly expressed by

$$C \sim \frac{1}{2}\mu b^2 . \tag{1.17}$$

As will be mentioned in Sect. 2.3, a dislocation moving at high speed experiences a friction force proportional to the velocity, Bv (B: the friction coefficient) due to scattering of phonons and electrons.

Summarizing, the equation of motion of a dislocation which is nearly parallel to the z-direction and gliding in the x-direction, as shown in Fig. 1.9, is given by

$$A\frac{\partial^2 x}{\partial t^2} + B\frac{\partial x}{\partial t} - C\frac{\partial^2 x}{\partial z^2} = \tau^* b . \tag{1.18}$$

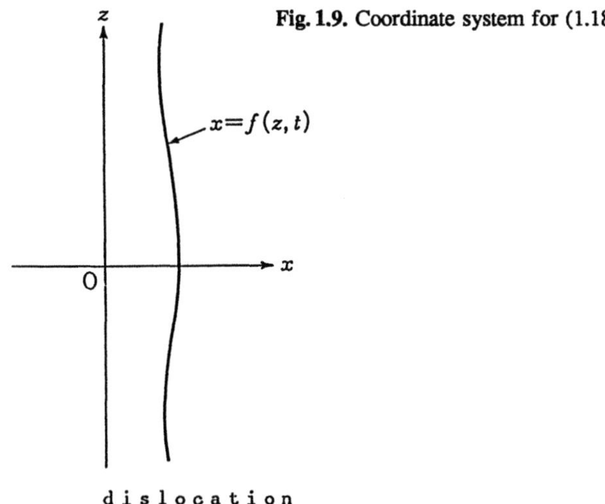

Fig. 1.9. Coordinate system for (1.18)

The first term is the inertia term, the second is the friction term, the third is the line-tension term, and the $\tau^* b$ on the right-hand side is the force acting on the dislocation due to the effective stress τ^*, which is the sum of the applied stress and the internal stress.

The problem of dislocation vibration can be solved by means of (1.18). The quasi-static motion of the dislocation can be treated by neglecting the first and the second terms. Figure 1.10 illustrates the glide process of a dislocation segment forming a network dislocation under the stress acting on the glide plane. A and B are the dislocation nodes that fix the two ends of the segment. With an increase in stress, the segment gradually bows out with a curvature determined by the balance equation (1.18). The maximum curvature is for the semicircular shape with a diameter of $\overline{AB}$ ($\equiv l$), position 2 in the figure. Thus, at a stress higher than τ_c, satisfying the balance equation for the semicircular state, i.e.

$$\tau_c = \frac{2C}{lb} \cong \frac{\mu b}{l} ,\tag{1.19}$$

the dislocation segment advances further to the positions $3 \rightarrow 4 \rightarrow 5$, turning completely around A and B, and eventually returns to its original position, emit-

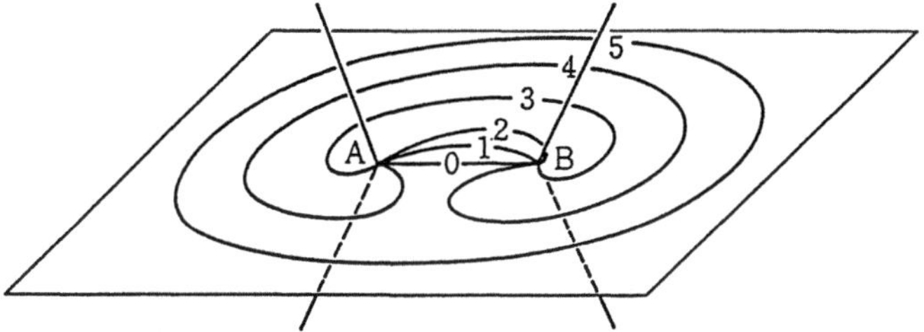

Fig. 1.10. Multiplication process of dislocation from a Frank-Read source. $0, 1, 2, \ldots$ indicate the sequence of the position of the dislocation segment

ting a dislocation loop. By repeating this procedure, many dislocation loops can be produced. This multiplication mechanism was proposed by Frank and Read, and has been verified experimentally by electron microscopy. Such a multiplication source is called the *Frank-Read source* and the critical stress expressed by (1.19) gives the source activation stress. In the case where strong obstacles such as precipitates are distributed on the glide plane with a spacing l, the critical stress at which a dislocation can glide through the obstacles is also given by (1.19). In this case, the τ_c of (1.19) is called the *Orowan stress*, after the researcher who first derived it.

1.5 Obstacles to Dislocation Motion

There exist various kinds of obstacles to the motion of dislocations. They control the flow stress of the crystal. These obstacles can be classified in different ways: (1) intrinsic obstacles to dislocation glide in an otherwise perfect crystal, and extrinsic obstacles due to the existence of lattice imperfections; (2) thermal obstacles, which can be surmounted by thermal activation, and athermal obstacles due to long-range interactions; (3) point obstacles, line obstacles, plane obstacles and bulk obstacles also exist. The main obstacles are listed below:

I. Intrinsic obstacles
   Friction (resistance to a high-velocity dislocation due to phonons and electrons)

   Peierls potential

II. Extrinsic obstacles
   Point obstacles (solute atoms, point defects)

   } Thermal obstacles

   Strong obstacles (dispersing particles, precipitates)

   Long-range stress field (other dislocations, coherent precipitates)

   } Athermal obstacles

2. Motion of Dislocations in Soft Metals

Motion of dislocations in fcc and hcp metals, where the Peierls force is very small as compared with other substances, is mainly impeded by obstacles such as impurity atoms or other dislocations distributed randomly. The motion is therefore quite random in nature. Local variations of the distribution of these obstacles may act as a strong impediment or conversely cause a discontinuous enhancement of motion. How the motion of dislocations settles into a steady state is the most important problem to be solved in order to understand the yield strength of these metals, because the yield stress is the minimum stress at which these metals deform plastically at a steady-state rate equal to the strain rate applied externally. Dislocations in such soft metals move at a speed of a few ms^{-1} or more at the yield point at room temperature. On account of such a rate of motion, dislocations are subject to strong frictional forces due to conduction electrons and lattice phonons. As a result, important and interesting behavior of dislocations characteristic of these metals is observed.

2.1 General Characteristics

An analysis that begins with the dislocation theory of the strength of metals and alloys has been thought to be the most useful in the accumulation of knowledge of the elementary processes accompanying gliding dislocations. This does not mean, however, that macroscopic plastic properties can be described by linear combinations of these elementary processes. Nonlinear couplings are in fact most important for understanding the strength of materials. This has begun to be recognized by researchers studying the plasticity of the so-called soft metallic materials, which include most metallic materials of fcc and hcp structures. In the following, we will first describe some predominant features of plastic deformation of these metals and alloys.

The slip system, i.e., slip planes and slip directions, are $\{111\}/\langle 1\bar{1}0\rangle$ for fcc metals and $\{0001\}/\langle 11\bar{2}0\rangle$ for hcp metals, respectively. Figure 2.1 shows a stereographic projection indicating the slip systems of fcc crystals. If the direction of tensile stress applied to a specimen crystal lies in triangle A with the three corners 111, 110 and 100, the primary slip plane is $(11\bar{1})$ and the slip direction [101]. The cross slip and conjugate slip systems with the largest Schmid factor are indicated by triangle B with the slip system $(1\bar{1}\bar{1})/[101]$ and C with $(1\bar{1}1)/[110]$,

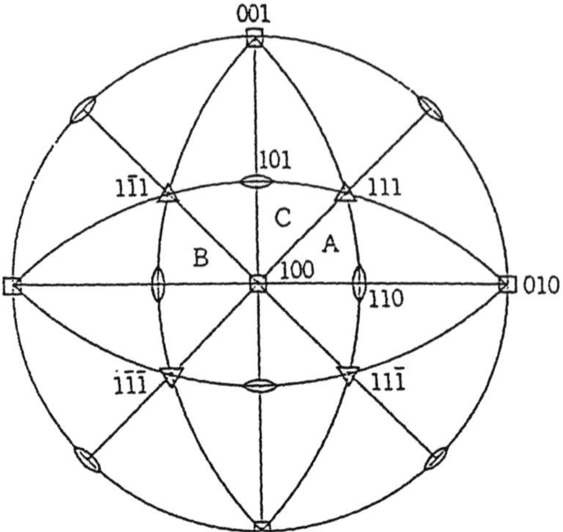

Fig. 2.1. Stereographic projection of slip systems of fcc crystals. When the axis of the applied tensile stress is in triangle A, the primary slip system, cross slip system, and conjugate slip system with the largest Schmid factors are illustrated. A: primary slip system $(11\bar{1})/[101]$
B: cross slip system $(1\bar{1}\bar{1})/[101]$ C: conjugate slip system $(1\bar{1}1)/[110]$

respectively. The Schmid factor is a geometrical factor for resolving the applied stress σ into the shear stress τ in the slip plane and along the slip direction concerned.

With increasing deformation by the operation of the primary slip system (A), the crystal axis oriented parallel to the tensile axis at the beginning of deformation rotates and then the conjugate slip system begins to operate, resulting in a double slip. However, the cross slip system has a common slip vector, the Burgers vector, parallel to [101], the same direction as the primary system, so that primary screw dislocations can change the slip plane into a cross slip plane. Such a change of slip plane is called a cross slip. The slip systems for crystals with various structures other than fcc and hcp are indicated in Table 5.1.

Figure 2.2 illustrates the relation between resolved shear stress (τ) and shear strain (γ) of metal crystals, copper (fcc), magnesium (hcp), and niobium (bcc), to show the structural characteristics revealed by plastic deformation behavior [2.1].

The most noticeable characteristics of fcc anf hcp metals are the smallness of flow stress at the beginning of the deformation and the minimal dependence of flow stress upon temperature. Stress-strain curves consist of the easy glide region (I) with the lowest hardening rate $\theta_I = (d\tau/d\gamma)_I$, the second stage or linear hardening stage (II) with the highest rate θ_{II}, and the final stage (III) with reversion to a small hardening rate θ_{III}. Although such behavior is common to metals with the above three kinds of crystal structures, fcc metals show the highest degree of hardenability. θ_I/μ for both fcc and hcp metals is almost 10^{-4}, where μ is the rigidity of the modulus. θ_I depends strongly upon the relative orientation of the axis of tensile stress to the crystal axis; that is, it is smallest

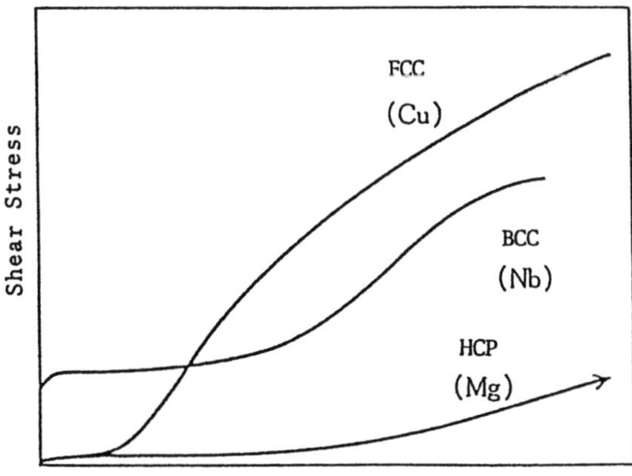

Fig. 2.2. Stress-strain curves for three typical metals of fcc, hcp, and bcc structures [2.1]

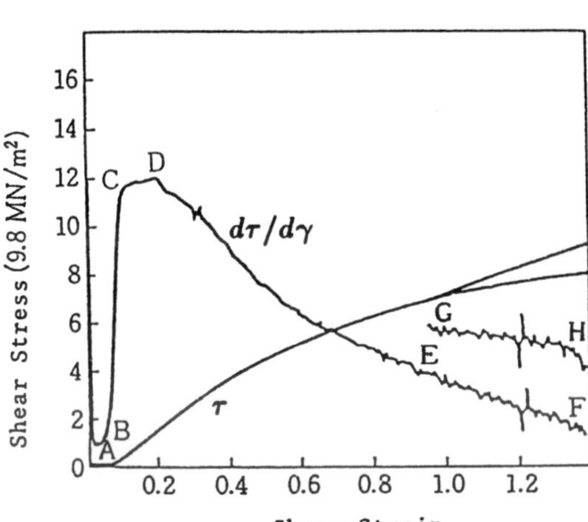

Fig. 2.3. Stress-strain and hardening rate-strain curves of a copper single crystal at 295 K. The strain rate is $2 \times 10^{-4}\,\mathrm{s}^{-1}$, and the axis of tensile stress is in triangle A in Fig. 2.1. The resolved shear stress is calculated as a function of the applied stress, taking into account the rotation of the crystal orientation with deformation, and then the hardening rate $d\tau/d\gamma$ is estimated. A indicates the easy glide region, BC the boundary between the easy glide and second glide regions, and CD the second glide region. The lower curves of τ vs γ and $d\tau/d\gamma$ vs γ for large γ are for where single glide deformation is assumed to occur, while the upper curves are for where the system is assumed to rotate along the boundary between the primary and conjugate slip systems on account of the double slip [2.2]

for the orientation inside the triangle (A) in Fig. 2.1 and highest for the ones near the three corners. It is also noticeable that θ_I of dilute alloys is generally smaller than that of pure metals. θ_{II}/μ for fcc metals and alloys is as large as $1/300$ as a result of hardening, due to the occurrence of secondary slip. θ_{II} does not depend much upon either the kind of metal or the solute concentration in dilute alloys, γ_I, the maximum strain for the first stage, in contrast, depends systematically upon the orientation, temperature, and alloy concentration. So far

15

we have considered the plastic deformation of single crystals. In polycrystalline materials it is noticeable that the above first stage of deformation does not appear; instead, the second stage starts at the beginning of deformation in order to provide continuity of stress and strain at grain boundaries by the introduction of multiple slip.

Figure 2.3 shows a typical set of data demonstrating the above-mentioned facts obtained on a copper crystal at 295 K by *Basinski* and *Basinski* [2.2].

2.2 Intrinsic and Extrinsic Barriers for the Motion of Dislocations

A dislocation in an otherwise perfect crystal can be activated by the application of exteral force larger than the resistive force resulting from the Peierls-Nabarro force. According to *Peierls* and *Nabarro* [2.3, 4], the resolved shear stress necessary to move an edge dislocation at 0 K is given by

$$\tau_P = \frac{2\mu}{1-\nu} \exp[-2\pi a/b(1-\nu)] , \qquad (2.1)$$

where τ_P is the Peierls stress, μ the rigity of the modulus, ν the Poisson ratio, a the distance between the lattice planes parallel to the slip plane, and b the magnitude of the Burgers vector. For fcc metals, a is the distance between $\{111\}$ planes equal to $(a_0/3)\langle 111 \rangle$, i.e., $a_0/\sqrt{3}$, and b is given by $(a_0/2)\langle 101 \rangle$, i.e. $a_0/\sqrt{2}$, where a_0 is the lattice constant. If $\nu = 1/3$, the above equation gives $\tau_P = 3.6 \times 10^{-6}\mu$. As seen from (2.1), τ_P depends strongly upon a/b. This is why metals with closely packed lattice structures exhibit the smallest τ_P. Table 5.1 lists τ_P experimentally observed for various materials. It is seen that the intrinsic Peierls-Nabarro barrier for the motion of dislocations in fcc and hcp metals is orders of magnitude smaller than for hard materials with other lattice structures.

For the reason mentioned above, the deformation of metals with fcc and hcp sturctures is primarily influenced by the existence of extrinsic defects such as dislocations and point defects. Figure 2.4 shows the critical shear stress τ_c of pure copper [2.5] plotted against the square root of the initial density of dislocations N_0. The proportionality of τ_c to $\sqrt{N_0}$ agrees satisfactorily with theories developed for the dislocation motion through dislocation forests [2.6]. However, τ_c of copper-nickel alloys [2.7–9] is independent of the dislocation density until the critical density characteristic of the nickel concentration is reached. In other words, τ_c is solely determined by the solute concentration in the case of a low density of initial dislocations. Even in copper of 99.999% purity, τ_c seems to become independent of N_0 below $5 \times 10^5 \, \mathrm{cm}^{-2}$, as shown in Fig. 2.4. The observed value of τ_c for pure copper, which has the lowest density of initial dislocations, is still larger than τ_P estimated by using (2.1), which may suggest that the motion of dislocations in pure copper is mainly controlled by impurity atoms of $\sim 10 \, \mathrm{ppm}$ concentration or some other defects acting as extrinsic obstacles.

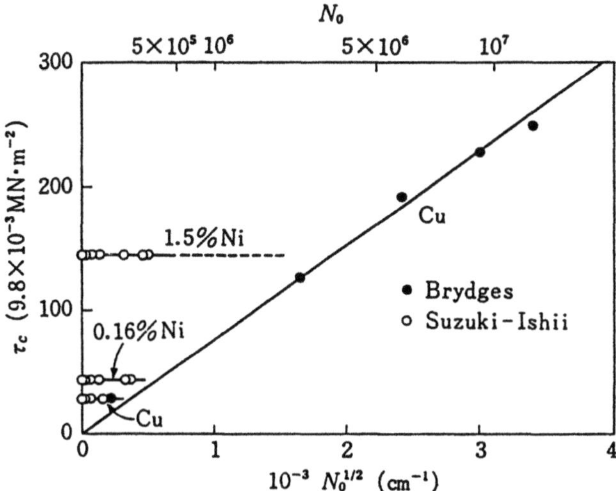

Fig. 2.4. Yield stress plotted against the initial dislocation density for pure copper [2.5] and for copper-nickel alloys [2.7–9]

Since dislocations in fcc and hcp metals can move at an average speed as high as 1m/s at the yield point, as will be shown later, quite different types of frictional forces, caused by interactions with conduction electrons and phonons act against the motion of dislocations. On account of these forces, remarkable phenomena occur in the plastic deformation of these metals, such as are rarely seen in other materials. This will be discussed later.

2.3 Dislocation Velocity

2.3.1 General

Johnston and *Gilman* [2.10, 11] measured the dislocation velocity in LiF crystals as a function of the applied stress by the etch-pit method. In LiF, as long as the crystal is pure enough, the motion of dislocations is governed by the Peierls mechanism. Dislocations can therefore move quite regularly compared with metals of low τ_P. So far most theories attempting to explain the yielding of crystals have studied mechanisms of generation of a number of glide dislocations (Frank-Read mechanism) and related phenomena such as locking of dislocation sources by impurities through elastic or chemical interaction. *Johnston* and *Gilman*, however, had great success in proving that yielding occurs when the steady-state flow governed by the following rule starts to occur:

$$\dot{\gamma}_{\mathrm{ext}} = bnv , \tag{2.2}$$

where $\dot{\gamma}_{\mathrm{ext}}$ is the externally imposed rate of deformation (i.e., the rate of deformation imposed by a testing machine), n is the mobile dislocation density, and v

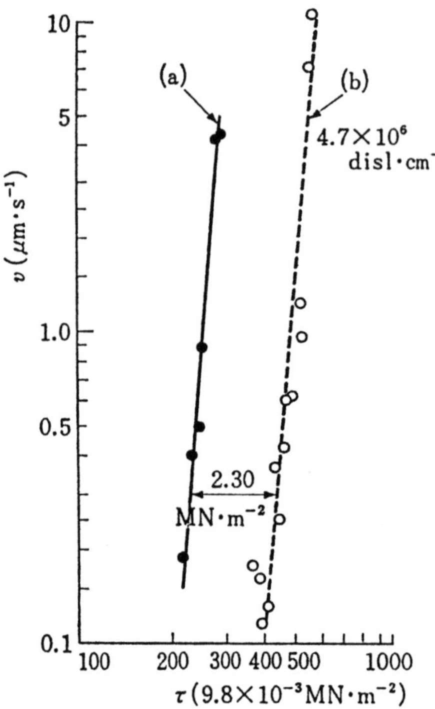

Fig. 2.5. Velocity of (screw) dislocation plotted against shear stress in LiF. Initial dislocation density of (a) $N_0 = 10^4\,cm^{-2}$, (b) $N_0 = 4.7 \times 10^6\,cm^{-2}$ [2.11]

is the average steady-state velocity, which is a function of applied shear stress τ. The relation holds in most metals. n can be taken to depend only implicitly upon τ. In other words, n is controlled by a kind of feedback mechanism operating between mobile dislocations and sources as will be discussed later.

The relation of v vs τ found in LiF is illustrated in Fig. 2.5a, where $N_0 = 1 \times 10^4\,cm^{-2}$. For a pulsed loading of stress, the dislocation motion is monotonic and regular, yielding a well-defined average velocity. Curve (b) shows v vs τ for the initial dislocation density increased to $4.7 \times 10^6\,cm^{-2}$ by a slight deformation of the specimen crystal used in (a). The two curves (a) and (b) are parallel to each other. If we write the effective stress acting on mobile dislocations as τ_e, it follows that

$$v(\tau_e) = v(\tau) \quad \text{and} \tag{2.3}$$

$$\tau = \tau_e + \tau_i \,, \tag{2.4}$$

where the internal stress $\tau_i = \alpha\sqrt{N_0}$ is found to be $2.30\,MN/m^2$ from Fig. 2.5.

The yield stress τ_c of pure copper is given by $\tau_e + \tau_i$ from the above equations, and τ_e should be constant at $\tau = \tau_c$ for different N_0 values, as illustrated in Fig. 2.4. It is also concluded that the density of mobile dislocations n should be constant independent of the magnitude of τ_c. *Pang* and *Galligan* [2.12] proved experimentally that n remained constant in the easy glide region, where τ_i increases with strain, in a Pb-In 5 at.% alloy crystal (Sect. 2.3.3).

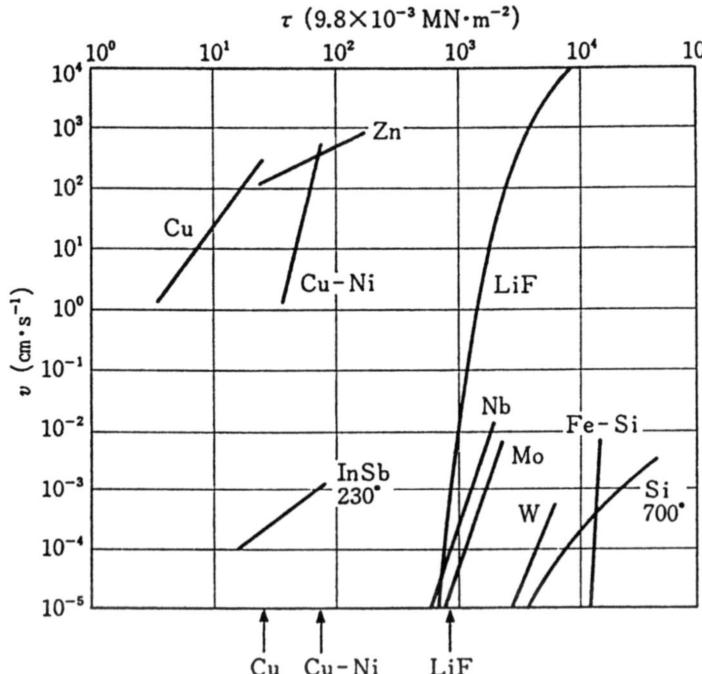

Fig. 2.6. Dislocation velocity as a function of shear stress in various materials measured at room temperature except for InSb and Si. Arrows indicate yield stresses of Cu, Cu-Ni and LiF, respectively

Figure 2.6 shows the dislocation velocity as a function of shear stress for various materials. It can be seen that dislocations in fcc and hcp metals move overwhelmingly fast compared to bcc metals and other nonmetallic materials. Figure 2.7a shows v vs τ for pure copper [2.13] and a copper-nickel alloy [2.7–9], and Fig. 2.7b is the plot for pure silver and silver-tin alloy [2.14], measured at 300 K and 77 K. The slopes of $\ln v$ vs $\ln \tau$ curves in (a) and (b) increase with alloying and with decreasing temperature. The motion of dislocations, therefore, is thought to be governed by thermal activation processes. Similar observations were made on copper alloys by *Haasen* and coworkers [2.15, 16]. In Fig. 2.6, the data on zinc were obtained under the application of high stresses by *Vreeland* and coworkers [2.17], in which the dislocation motion is possibly controlled by the frictional force caused by the interaction with phonons. The velocity of dislocations can be seen to approach the sound velocity with increasing applied stress. The difference in the velocity of edge and screw dislocations is not large in fcc metals, but in bcc metals and in semiconductors with the diamond or zinc blende structures. For example, screw dislocations move at a speed of 1/10 of that of edge dislocations in iron at 153 K [2.18].

The dislocation velocities at yield points and yield stresses for various materials are listed in Table 2.1. Remarkable features of fcc and hcp metals can be seen from this table. The ductility characteristic of these metals is, needless to say,

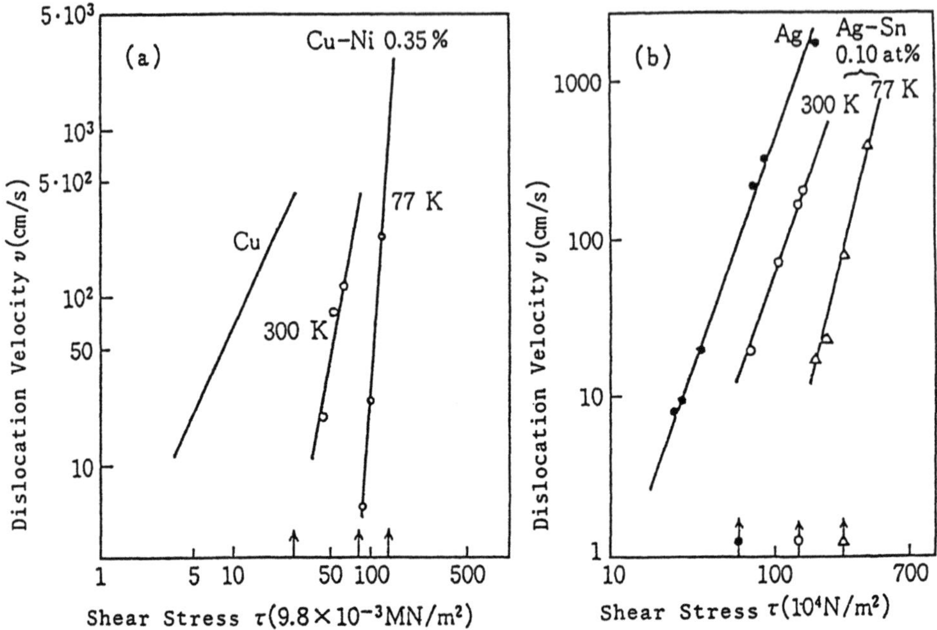

Fig. 2.7. (a) Dislocation velocity as a function of shear stress for pure copper and copper-nickel 0.35 at.% alloy [2.7–9, 13]. (b) Dislocation velocity as a function of shear stress for pure silver and silver-tin 0.10 at.% alloy [2.14]. Arrows along the abscissa indicate the yield stresses of the various metals and alloys. Solid lines for the Cu-Ni alloy are calculated curves, see Sect. 3.3

Table 2.1. Dislocation velocity at the yield point for various materials

Material	Temperature [K]	Yield stress $[\text{MN m}^{-2}]$	Dislocation velocity $[\text{cm s}^{-1}]$	Ref.
Si	873	327.3	1×10^{-4}	2.19
LiF	300	8.8	1×10^{-3}	2.11
W	300	186.2	0.07	2.20
Fe-Si 3.35%	298	137.2	4×10^{-5}	2.21
Cu	298	0.25	8×10^2	2.13
	77	0.25	8×10^2	2.13
Cu-Ni 0.35%	298	0.78	6×10^2	2.7, 8
	77	1.31	3×10^3	2.7, 8
Ag	300	0.60	1×10^2	2.14
Ag-In 0.5%	300	2.00	2.9×10^2	2.14

due to the high mobility of dislocations. The details of the ductility of these metals are not discussed here, because it is necessary to discuss work-hardening and related phenomena in more detail, which will be done in Chaps. 8 and 9. It is just mentioned that the above soft metals possess the largest work-hardenability and ductility on account of the fact that glide dislocations belonging to the primary or the secondary system can be activated with equal ease in the work-hardened state to reduce the inhomogeneity compared with other substances.

2.3.2 Characteristics of Dislocation Motion in fcc and hcp Metals

Characteristics of the motion of dislocations in fcc and hcp metals, as long as the basal slip is involved in the latter, are quite different from those in LiF. Figure 2.8 shows an etch-pits photograph [2.13] taken on copper loaded by pulsed stresses, the first pulse being equal to $0.116\,\mathrm{MN\,m^{-2}}$, and the second one, $0.191\,\mathrm{MN\,m^{-2}}$. The surface of the crystal was etched before and after every loading of the pulse stresses. We can see from this photograph the irregularity of the motion of the dislocations. Figure 2.9 shows histograms of the number of moved dislocations versus the distance of movement obtained for pure copper (a) and for Cu-Ni 0.16 at.% alloy (b) [2.7]. A similar histogram was obtained for silver alloys [2.14]. Except for a small number of dislocations that moved over the maximum distances, most dislocations did not move with the application of a second stress pulse of the same magnitude as the first one. They were undoubtly obstructed by strong barriers, which stopped the first movements as well. It should be noticed that Fig. 2.7 was obtained from the maximum distances of movements, that is, illustrates v_{max} vs τ. Whether these v_{max} values really give the velocities corresponding to the steady-state deformation rate expressed by (2.2), and what the strong barriers are, are most important subjects to be studied.

Computer simulations of dislocation motion through a random distribution of point obstacles were first worked out by *Foreman* and *Makin* [2.22]. Their simulations were based on the principle that a critical line tension overcomes the impeding force due to a point obstacle without thermal activation under a given shear stress, i.e., it holds at $0\,\mathrm{K}$. Their result is shown in Fig. 2.10. The number of point obstacles of the same strength is 100, which are distributed randomly. A dislocation initially at I moves to the position FSI with increasing stress. Thereafter it moves to the final position without increasing stress, that is, yielding occurs at FSI. It is noticeable that even after the greater part of the dislocation line has reached the final position F, the line near the surfaces does not

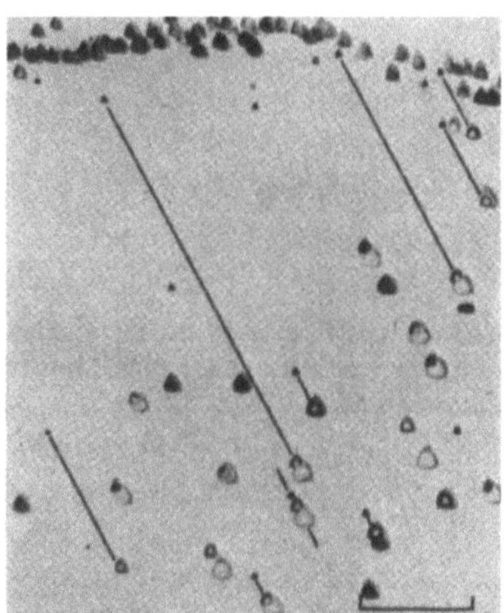

Fig. 2.8. Dislocation motion in pure copper studied by the etch-pit method. A first pulse of stress equal to $0.116\,\mathrm{MN\,m^{-2}}$ for 1s and a second, equal to $0.191\,\mathrm{MN\,m^{-2}}$ for 1s were applied. The scale in the photograph is $100\,\mu\mathrm{m}$ [2.13]. Pits of the largest, medium, and smallest sizes correspond to the positions of dislocations before loading, after a first pulse of stress and after a second, respectively

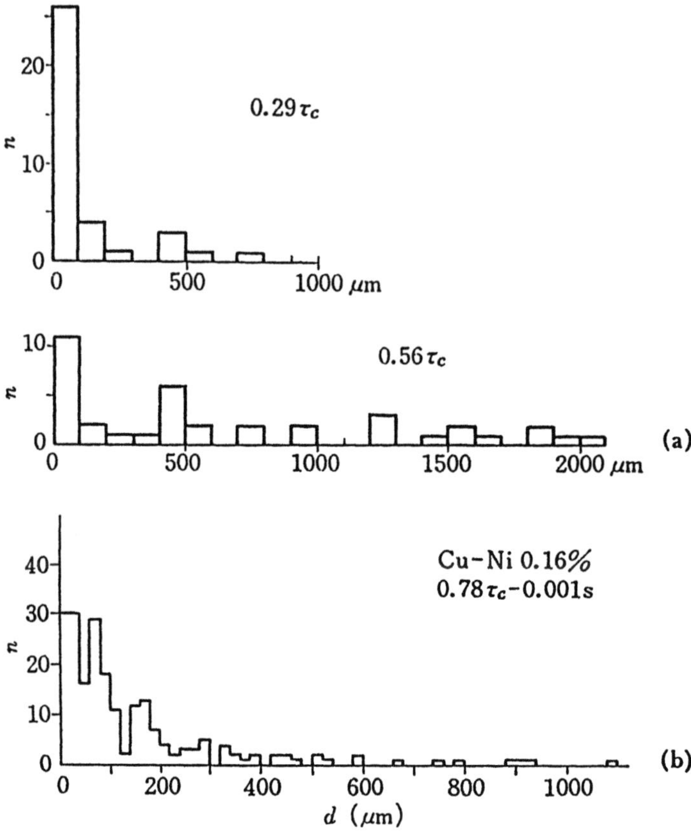

Fig. 2.9a,b. Histograms of the number of moved dislocations vs the distance of movement. (a) Pure copper: the applied square-shaped pulse of stress is equal to $0.29\tau_c$ and $0.56\tau_c$ for 5×10^{-4} s, respectively. (b) Cu-Ni 0.16 at.% alloy: the applied pulse of stress with a similar shape as in (a) is equal to $0.78\tau_c$ for 10^{-3} s. The measurements were carried out at room temperature [2.7]

move as much (as illustrated by A or B). The etch-pit method is inherently one to observe the surface of a specimen crystal. The irregularities of a dislocation motion as suggested by Fig. 2.9 may therefore be doubtful, because the inner parts of the dislocations may move much larger distances than those expected from the observations of etch pits. In order to clarify such a doubt it is necessary to carry similar simulations of dislocation motion for a larger number of point obstacles and at finite temperatures.

With respect to the steady state of dislocation motion at finite temperatures in the field of a random distribution of point obstacles, computer simulation studies carried out by *Wynblatt* [2.23] are of great interest. Suppose a dislocation under a shear stress, after passing over a number of barriers without the assistance of thermal fluctuations, meets a barrier consisting of an array of point obstacles that cannot be overcome only with the aid of the applied stress. After surmounting this barrier with the aid of thermal activation, the dislocation meets the next

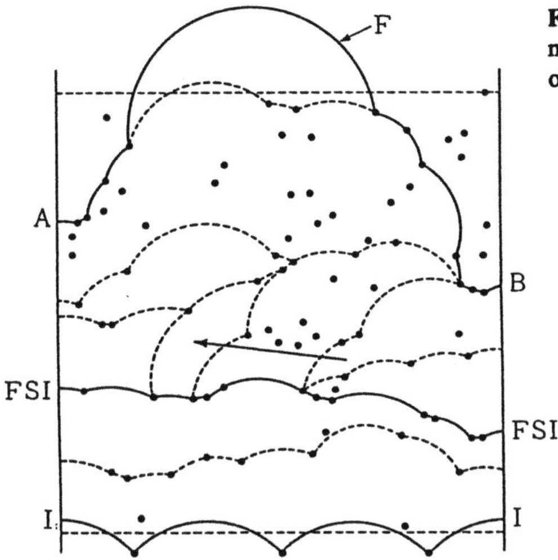

Fig. 2.10. Computer simulation of dislocation motion through a random distribution of point obstacles [2.22]

barrier, and so on. In Fig. 2.11, the average strain rate for the jth activation process denoted by $\dot{G}^j$ is plotted against j, the number of processes. Taking A^j to be the area swept by the dislocation after the j-th process, p^j the average rate per point obstacle of surmounting the jth barrier, $\langle l \rangle^j$, the average distance between point obstacles along the dislocation line, to be equal to $\langle l \rangle$, which is independent of j, we have

$$\dot{G}^j = A^j p^j / \langle l \rangle^2 . \tag{2.5}$$

As seen from Fig. 2.11, $\dot{G}^j$ varies greatly with j. The deformation does not seem to proceed at a steady-state rate with a constant activation energy. This situation is very similar to that of the dislocations we met in the etch-pit experiment

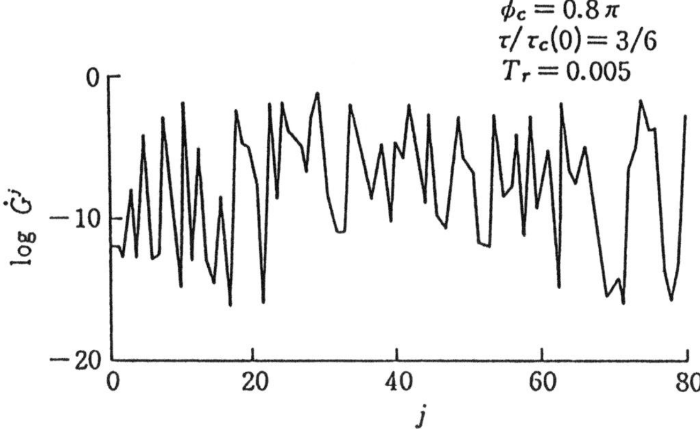

Fig. 2.11. Average strain rate for the jth activation process plotted against the process number. T_r indicated in the figure is the reduced temperature given by $k_B T/(1/2)\mu b^3$, ϕ_c is the measure of the strength of a point obstacle given by (3.8), and $\tau_c(0)$ is the yield stress at 0 K [2.23]

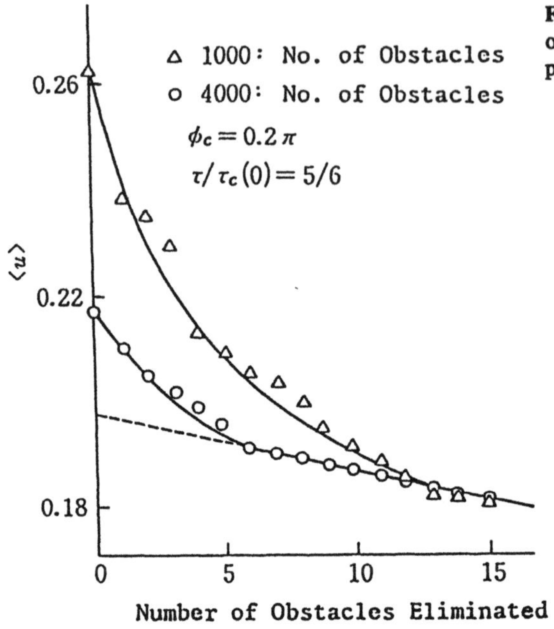

Fig. 2.12. Average activation energy per obstacle plotted against the number of processes eliminated [2.23]

described before (Fig. 2.9). In other words, there are a number of barriers in the path of a dislocation that cannot be overcome by the repeated application of stress pulses of the same magnitude, as *Wynblatt* showed by computer simulations.

Let us eliminate some of the strong barriers in Fig. 2.11. The result is shown in Fig. 2.12, in which the abscissa is the number of strong barriers eliminated. As seen from this figure, the average thermal activation energy per barrier $\langle u \rangle$ tends to converge, with the elimination of a relatively small number of the strongest barriers, to a constant value independent of the number of point defects used for simulation. The procedure of determining the rate of deformation in the steady state by inserting v_{max} into (2.2) can be regarded as the same in principle as the one for determining the activation energy for the steady-state motion of dislocations by eliminating a number of strong barriers in the computer simulation. It can be concluded that a relatively small number of dislocations are in effect responsible for the steady-state flow rate under a constant shear stress.

2.3.3 The Steady-State Velocity and Number of Moving Dislocations

Equation (2.2) is the fundamental equation giving the relation between the steady-state velocity of dislocations derived from the microscopic motion of dislocations and the macroscopic strain rate of a specimen crystal. The microscopic velocity of dislocations is of course influenced by the above-mentioned strong barriers in their path. Accordingly, we have first to explain the mechanism that gives rise to a steady-state flow irrespective of the distribution of barriers varying in strength.

We will explain here how the effect of strong barriers is absorbed into the average velocity of dislocations. Suppose that strong barriers of equal strength are distributed at an average distance equal to L along the direction of motion of dislocations, and take the time spent at one of these barriers before surmounting it to be t_{sb}. Of course, L is much larger than the average separation distance of point obstacles l. Now, the average velocity of dislocations, taking the time needed to move between neighboring strong barriers as t, will be given by

$$\langle v \rangle = L/(t + t_{sb}) = (L/t)\{1/[1 + (t_{sb}/t)]\} , \tag{2.6}$$

which can be rewritten as

$$\langle v \rangle = \alpha v_{max} \quad \text{and} \quad \alpha = 1/[1 + (t_{sb}/t)] . \tag{2.7}$$

For the motion of a single dislocation as in the computer simulation, this relation will give $\alpha < 1$, and, accordingly, t_{sb} cannot be neglected. If a number of dislocations are generated from the sources, however, most of them will pile up in front of the strong barriers with time, since $v_{max} \gg v_{sb} = l/t_{sb}$, where v_{sb} is the local velocity at the strong barrier. Taking the number of piled-up dislocations as m, the shear stress precipitating the leading dislocation of this piled-up group becomes m times as large as the applied shear stress [2.24]. As a result, t_{sb} will decrease until it becomes $v_{sb} = v_{max} = \langle v \rangle$. The value of m must vary in accordance with the strength of the barrier to give the relation $v_{sb} = \langle v \rangle$ under constant applied shear stress, so that the steady-state motion of dislocations is the result. At the same time, however, it must be remembered that the density of dislocations increases because of piling in association with establishing a steady-state flow, which gives rise to τ_i as expressed in (2.4).

The density of dislocations found in the first stage of deformation of copper is known experimentally to increase as much as $\sim 10^6\,\mathrm{cm}^{-2}$ from an initial density of $10^4\,\mathrm{cm}^{-2}$. Using (2.2), the number of mobile dislocations can be estimated as $40\,\mathrm{cm}^{-2}$ for given values such as $\dot{\gamma}_{ext} = 10^{-4}\,\mathrm{s}^{-1}$ and $v = 10\,\mathrm{cm/s}$. This suggests that only a small proportion of the total dislocation density contributes to a steady-state flow, as expected, and most of the initial dislocations contribute to an increase in overall density of dislocations piled up before a number of strong barriers.

Recently, *Galigan* [2.25] successfully measured the number of mobile dislocations as a function of shear strain by observing the variation of the magnetic flux density during deformation at a constant rate in the mixed state of a lead-indium superconducting alloy.[1] As seen in Fig. 2.13, the number of mobile dislocations n, proportional to the change of the magnetic flux density in the

---

[1] In the mixed state of a superconductor of the second kind, magnetic flux penetrates into a sample crystal to form a lattice of fluxoids. This flux-line lattice is partly stabilized in the presence of an external magnetic field by defects such as dislocations. When the defects move, a change in the flux density will occur, which can be measured by an ac bridge technique. *Pang* and *Galligan* showed that the change of the flux density was proportional to the number of mobile dislocations.

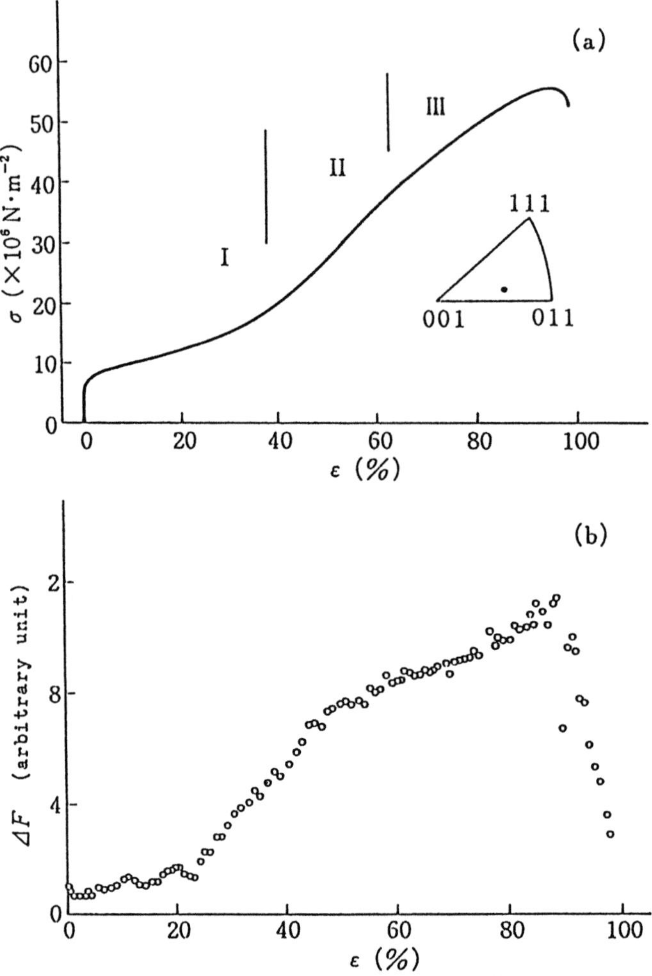

Fig. 2.13. (a) Stress-strain curve for lead-indium 5 at.% alloy measured at a strain rate of $10^{-4}\,\mathrm{s}^{-1}$ at 4.2 K. (b) The change in flux density accompanying the motion of dislocations as a function of shear strain [2.25]

easy glide region, where shear strain is less than 20%, is almost constant. In the second stage of deformation, where shear strain is greater than 20%, however, the number of mobile dislocations increases steeply. The fundamental equation (2.2) does not hold in this stage in the same sense as in the initial stage of deformation, because the mode of deformation changes on account of the formation of new barriers to the motion of dislocations, such as dislocation forests or cell boundaries (Chap. 9).

2.4 Frictional Forces
due to Conduction Electrons and Phonons

When the frictional force acts on a dislocation moving at a velocity v, it is given by

$$Bv = b\tau \, , \tag{2.8}$$

where $b\tau$ is the component of the external shear force acting in the direction of motion of the dislocation, which is in equilibrium with the frictional force Bv. In general, B is given by the sum of the frictional coefficient B_e due to conduction electrons and B_p due to phonons:[2]

$$B = B_e + B_p \, . \tag{2.9}$$

As a result, dislocations dissipate energy when forced to oscillate by an ultrasonic wave. The corresponding attenuation α_D was described by *Granato* and *Lücke* [2.26, 27], applying a string model to vibrating dislocations, as

$$\alpha_D = \int_0^\infty \alpha(l)N(l)\, l dl \, , \tag{2.10}$$

where $\alpha(l)$, the attenuation per unit length of dislocation, is given by

$$\alpha(l) = \frac{4R\mu b^2}{\pi^2 A v_s} \frac{\omega^2 d}{(\omega_0^2 - \omega^2)^2 + (\omega d)^2} \, , \tag{2.11}$$

where $d = B/A$ with $A = \pi\varrho b^2$, ϱ is the density of the material, ω the frequency of the applied ultrasonic wave, μ the shear modulus, R the orientation factor, and v_s the shear wave velocity. The fundamental frequency of a dislocation of length l designated by ω_0 is

$$\omega_0 = (\pi/l)(C/A)^{1/2} \, , \tag{2.12}$$

where C is the line tension equal to $\mu b^2/2$. The number of dislocations $N(l)$ whose loop length lies between l and $l + dl$, is assumed to be given by

$$N(l)dl = (\Lambda/L_c^2) \exp(-l/L_c)dl \, , \tag{2.13}$$

where Λ is the total length of dislocation per unit volume, and L_c is the average loop length of the dislocations.

In order to determine α_D by means of the ultrasonic attenuation experiment, it is necessary to evaluate Λ or L_c by the etch-pit method or otherwise. This

---

[2] For the vibrational motion of a dislocation, it is accompanied by the third frictional coefficient B_r due to the bremsstrahlung. B_r is given by $1/8\varrho b^2\omega$, where ϱ is the density of the material and ω the vibrational frequency of the dislocation. B_r is estimated to be a few tenths of B_e and it is often neglected in the following discussion.

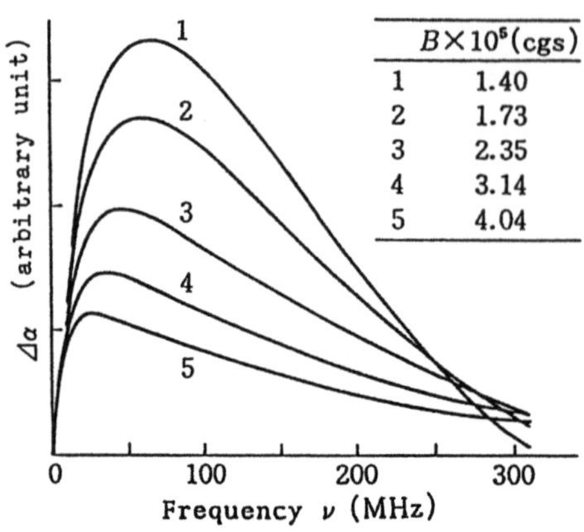

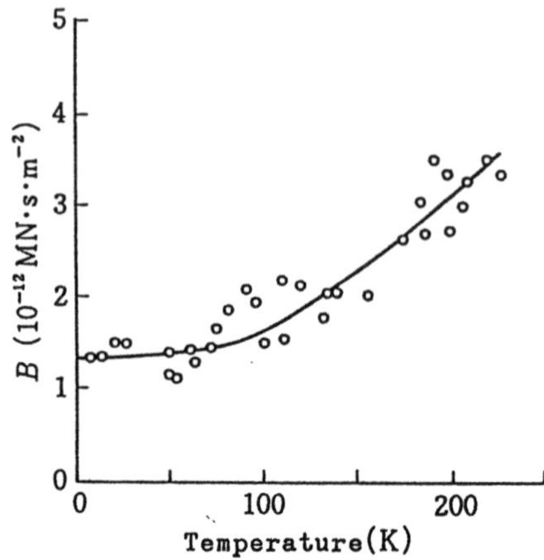

Fig. 2.14. Calculated curves of $\Delta\alpha_{\mathrm{D}}$ vs ν [2.28]

Fig. 2.15. Temperature dependence of B in aluminum [2.28]

introduces an uncertainty to some extent into the evaluation of B. In order to overcome the above uncertainty, *Hikata* et al. [2.28] devised a new technique to measure the incremental attenuation $\Delta\alpha_{\mathrm{D}}$ as a function of frequency by applying a small dynamic bias stress wave. From an analysis of this incremental attenuation it is possible to extract the value of B without knowing the dislocation density. Since it is expected that the background attenuation is not affected by this small bias stress, the attenuation change obtainable in this method is solely due to the increase in the average loop length of dislocations. As illustrated in Fig. 2.14, the $\Delta\alpha_{\mathrm{D}}$ vs ω plot thus obtained is matched with the theoretical relation:

$$\Delta\alpha_{\mathrm{D}} = \frac{\partial}{\partial L_{\mathrm{c}}} \int_0^\infty \alpha(l) l N(l) dl (\delta L_{\mathrm{c}})$$

$$= \frac{\Lambda}{L_{\mathrm{c}}^3}(\delta L_{\mathrm{c}}) \int_0^\infty \alpha(l) l \left(\frac{l}{L_{\mathrm{c}}} - 2\right) \exp\left(-\frac{l}{L_{\mathrm{c}}}\right) dl \; . \tag{2.14}$$

Since Λ and L_{c} in (2.14) contribute as parameters outside the integral, the forms of $\Delta\alpha_{\mathrm{D}}$-$\omega$ curves are completely determined by the resonant frequency $\nu_{\mathrm{D}} = \omega_0/2\pi$ and B.

Using the above experimental method, *Hikata* obtained B in aluminum as a function of temperature as shown in Fig. 2.15. Below 40 K, B is independent of temperature, and above 40 K, it increases with temperature. According to theory, as explained later, B_{e} takes a constant value independent of temperature, while B_{p} increases with increasing number of phonons in the low temperature region and becomes linear with temperature in the region $T > \theta_{\mathrm{D}}$, where θ_{D} is the Debye temperature. In lead B_{e} behaves similarly to that in aluminum [2.29]. Table 2.2 shows B for various materials measured by the ultrasonic method. Data without indication of temperature were all measured at room temperature.

Table 2.2. Friction coefficient B for various materials measured by the ultrasonic method

Material	$B \times 10^4$ (cgs)	Ref.
LiF	2.5	2.30
	2.4	2.31
NaCl	2.5–10.5	2.31
	1.6	2.32
KCl	3.2	2.33
Al	10.0	2.34
	0.14 (20–40 K)	2.28
Cu	8.0	2.35
	6.5	2.36
	1.2	2.33
Pb	3.7	2.34
	1.1 (60 K)	2.34
	0.86 (20–40 K)	2.29

2.5 Theoretical Studies of the Frictional Forces

2.5.1 Frictional Force due to Conduction Electrons

The interaction of moving dislocations with conduction electrons is attributed to the interaction of propagating elastic waves with the latter. After consideration of many complications, the conclusion that B_e does not depend upon temperature is reached by *Holstein* [2.37], *Kravchenko* [2.38] and *Brailsford* [2.39]. *Holstein* obtained B_e from standard perturbation theory, while *Kravchenko* and *Brailsford* on the basis of the Boltzmann equation. Here we will describe the outline of *Brailsford*'s theory.

If the displacement field of a stationary dislocation is $w(r)$, the corresponding form for a dislocation moving with uniform velocity v_d is $w(r - v_d t)$, provided v_d is small compared with the sound velocity. The local lattice velocity $u(r, t)$ is then given by

$$u(r, t) = \sum_q u_q \exp[iq \cdot (r - v_d t)] \,, \tag{2.15}$$

where $u_q = \sum_\lambda -iq \cdot v_d e_{q\lambda} w_{q\lambda}$. Here q is the wave number of the lattice wave, λ an index specifying the normal mode with polarization vector $e_{q\lambda}$, and $w_{q\lambda}$ the Fourier amplitude of the displacement associated with the mode. Taking the power dissipated by each component of (2.15) to be P_q, as a result of the motion

$$P_q = A_\parallel |u_{q\parallel}|^2 + A_\perp |u_{q\perp}|^2 \,, \tag{2.16}$$

where $\parallel$ and $\perp$ refer to longitudinal and transverse components, respectively. As the effective drag force per unit length of dislocation is given by $B_e v_d$, it gives

$$B_e = \frac{1}{L v_d^2} \sum_q P_q \,, \tag{2.17}$$

where L is the length of dislocation. The problem is thus reduced to the theoretical evaluation of $A_\parallel$ and $A_\perp$ in (2.16). Fortunately, their electronic components have been calculated [2.40] and so one only needs reiterate the pertinent solution. After calculation, it is found that $A_\perp \ll A_\parallel$. Thus, for an edge dislocation, the longitudinal-wave contribution to B_e, independent of temperature, is

$$B_e = \left(\frac{1 - 2\nu}{1 - \nu}\right)^2 \frac{n_0 m v_F b^2 q_D}{96} \phi(q_D/q_{TF}) , \qquad (2.18)$$

where $\phi(x) = (1/2)[(1 + x^2)^{-1} + x^{-1} \tan^{-1} x]$. In the above equation, n_0 is the equilibrium value of the electron density, m the mass of the free-electron, v_F the Fermi velocity, q_D the radius of a Debye sphere, q_{TF} the reciprocal of the Thomas-Fermi screening length, and ν Poisson's ratio. Meanwhile, for a screw dislocation the longitudinal-wave contribution to B_e is taken to be zero.

Apart from numerical factors, (2.18) agrees with the result obtained by *Holstein*. Thus, both approaches led to a temperature-independent electronic contribution to the drag force.

2.5.2 Frictional Force due to Phonons

A dislocation moving in the flux of phonons experiences a frictional force. Two mechanisms of the dislocation-phonon interaction have been considered so far. One is the nonlinearity mechanism (anharmonicity of elastic field) [2.41–44] and the other is the fluttering mechanism [2.45–50]. In the fluttering mechanism, a dislocation in translational motion interacts with phonons via inelastic scattering accompanying the momentum transfer, which gives rise to the inertia for the translational motion, that is, the retarding force.

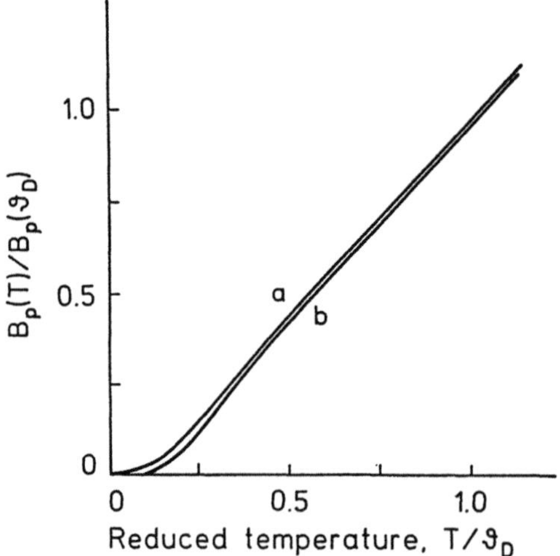

Fig. 2.16. Calculated curves of B_p vs T/θ_D. (*a*) Fluttering mechanism [2.45–47] and (*b*) Nonlinearity mechanism [2.43]

According to *Ninomiya*, the frictional coefficient due to phonons is given by

$$B_\mathrm{p} = \frac{k_\mathrm{B} T \omega_\mathrm{D}^2}{\pi^2 c^3} \, , \qquad \text{(high temperatures)} \, ,$$

$$B_\mathrm{p} = \frac{14.4 k_\mathrm{B} T \omega_\mathrm{D}^2}{\pi^2 c^3} \left(\frac{T}{\theta_\mathrm{D}} \right)^2 \, , \qquad \text{(low temperatures)} \, ,$$

$$(2.19)$$

where c is the sound velocity, ω_D the Debye frequency, and θ_D the Debye temperature. The magnitude of B_p at 300 K is about 1.6×10^{-4} for copper, which agrees with the ultrasonic measurement [2.33] listed in Table 2.2. In Fig. 2.16, $B_\mathrm{p}(T)/B_\mathrm{p}(\theta_\mathrm{D})$ is plotted against temperature. The fluttering mechanism (*Ninoiya*) and the nonlinearity mechanism (*Brailsford*) gives almost equal B_p except for at low temperatures. In the low temperature region, B_p shows T^3 dependence in the former, but T^5 dependence in the latter.

3. Dislocation Motion in the Field of a Random Distribution of Point Obstacles: Solution Hardening

Deformation rates of crystals, according to many experimental and theoretical studies, are controlled by thermal activation processes for surmounting various barriers. So far, the problems have been discussed using the following implicit solutions: (1) the processes are quasi-static and do not involve any dynamical aspects; (2) the frictional forces produced in materials are neither extremely large nor small. If the frictional forces due to the material are very small, the hypothesis (1) does not hold any more. This is the case of deformation of superconductors as described in the next chapter. *Weertman* [3.1] dissussed creep of ice on the basis of visco-elastic deformation theory developed by *Eshelby* [3.2]. According to ultrasonic attenuation experiments [3.3], B is 10^7 times as large as that for metals. In this case, although the frictional force is still proportional to the velocity of dislocations, the time of motion in the area between barriers becomes longer than that needed to surmount each of them. In other words, the barriers are of no importance for the discussion of the rate of deformation. Accordingly, such a linear visco-elastic deformation may be put aside in the present discussion. As for studies of plastic deformation of crystalline materials, the past treatments of the case, where the frictional forces are very small, should be carefully reconsidered. In the present chapter, we discuss problems within the above framework, (1) and (2), and those outside of this will be discussed in the next chapter.

3.1 Solution Hardening

Flow stress, which causes the steady-state motion of dislocations through random arrays of solute atoms in a solid solution, is in general larger than in pure metals. This phenomenon is called solid-solution hardening. Here we are mainly concerned with dilute alloys of fcc structure.

3.1.1 Experimental

Figures 3.1 and 3.2 show τ_c vs T and τ_c vs c for Cu-Ni alloys [3.4, 5], while Figs. 3.3 and 3.4 show these for Ag-Al alloys [3.6]. Figure 3.5 illustrates the details of τ_c vs T at low temperatures for a Cu-Ni alloy. In Cu-Ni alloys, τ_c is proportional to $c^{1/2}$ in the range of concentration $c < 2$ at.%, and proportional to c for $c > 2$ at.%. In Ag-Al alloys, however, τ_c is proportional to $c^{1/2}$ in the whole

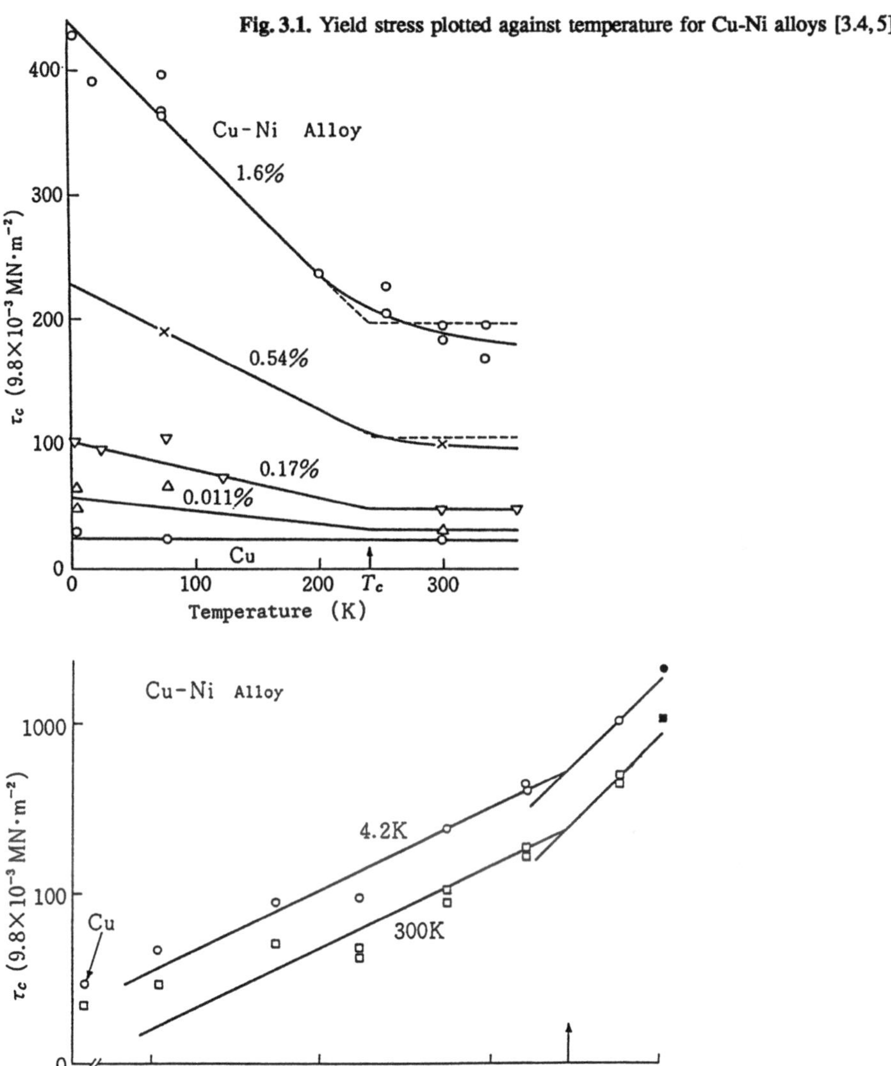

Fig. 3.1. Yield stress plotted against temperature for Cu-Ni alloys [3.4,5]

Fig. 3.2. Yield stress plotted against the concentration of nickel for Cu-Ni alloys [3.4,5]

range of concentration studied. As for the temperature dependence, although it is obscure on account of the anomalous behavior of τ_c at low temperatures, it is judged as $\tau_c \propto T$ for Cu alloys and $\tau_c \propto T^{2/3}$ for Ag alloys except for the anomaly.

The anomalous behavior of τ_c below 100 K, as shown by the dotted curve in Fig. 3.5, is a quite general one in dilute alloys of fcc structure such as Cu-Ag [3.7], Cu-Al [3.8], Cu-Ge [3.9], Cu-Co [3.10], Ag-Sn [3.11], Al-Mg [3.12], and others. We will come back to this subject later.

33

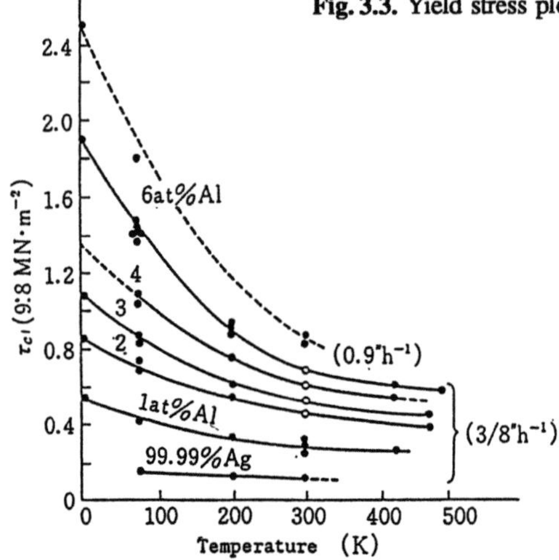

Fig. 3.3. Yield stress plotted against temperature for Ag-Al alloys [3.6]

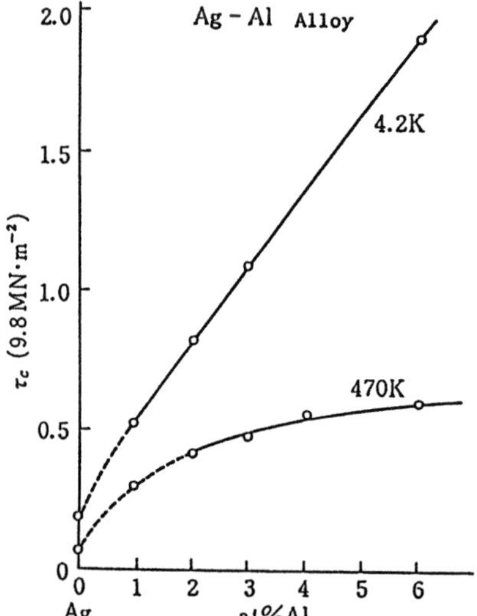

Fig. 3.4. Yield stress plotted against the concentration of aluminum for Ag-Al alloys [3.4]

One of the characteristics of the solid-solution hardening of fcc alloys is the dependence of τ_c with respect to temperature above 300 K, which was studied in detail for many alloys by *Haasen* and coworkers [3.13–15] and *Kloske* and *Fine* [3.16]. Writing τ_c in this range of temperatures as τ_0, which is called the plateau stress, we can plot τ_0 as in Fig. 3.6 as a function of $c^{1/2}$ for various dilute alloys of gold [3.13, 14].[1]

---

[1] *Haasen* et al. plotted τ_0 against $c^{2/3}$ in their original papers. We have replotted their data against $c^{1/2}$, however. Which of the two c dependences of τ_0 is to be preferred can hardly be determined from their data.

34

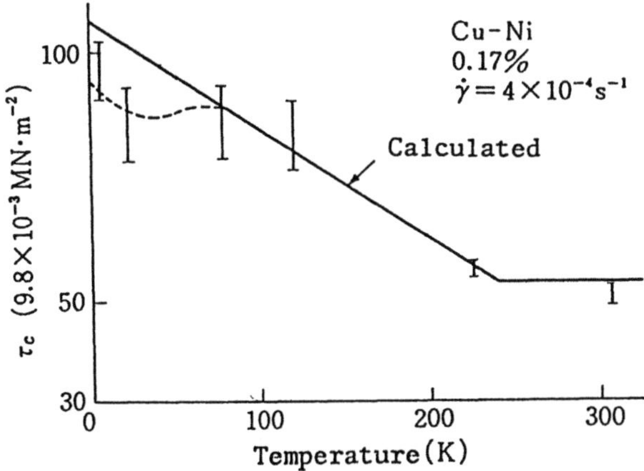

Fig. 3.5. Yield stress plotted against temperature for Cu-Ni 0.17 at.% alloy [3.4, 5]. Anomalous behavior of yield stress at low temperatures, below 100 K, is due to the decrease of the phonon frictional force. Experimental points were obtained on different specimens cut from a large single crystal. Solid line is a calculated curve, see Sect. 3.3

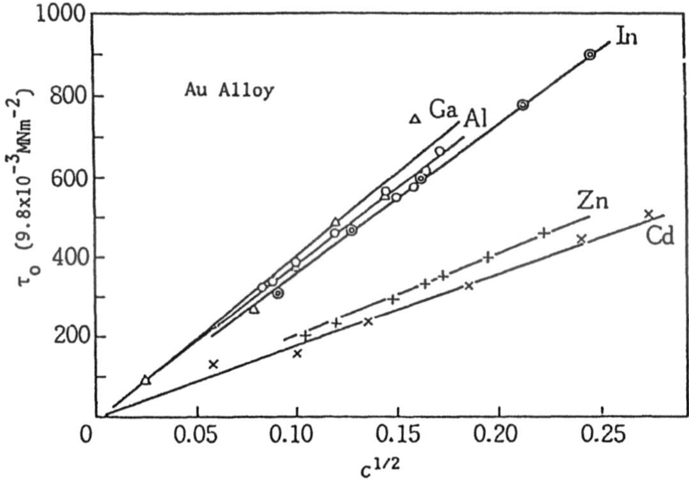

Fig. 3.6. Plateau stress as a function of the concentration of solute atoms for gold alloys. See footnote 1. [3.13, 14]

3.1.2 Theoretical

A dislocation in a fcc crystal is usually extended into two partial dislocations separated by a stacking fault of width l. In dilute alloys, e.g., 1 at.%, the average distance between solute atoms is on the order of l, which is equal to $10\,b$ or so for Cu alloys. Hence, as long as the elastic interaction between dislocation and solute atoms is important, it is allowable to treat an extended dislocation as a perfect dislocation [3.17]. From the standpoint emphasizing chemical interaction of solute atoms with an extended dislocation via a stacking fault, however, the above approximation is not allowed. The theory based on the chemical interaction [3.18] gives a different dependence of τ_c upon c from experimental results described in Sect. 3.1.1, and, besides, the hardening depends not upon the stacking-fault energy but upon the elastic properties of solute atoms embedded in the solvent lattices. In general, therefore, most theories of solution hardening disregard the chemical interaction of solute atoms with the dislocations.

There are two kinds of elastic interactions of dislocations with solute atoms, (i) interaction via the elastic strain field of the solute atom (size effect) and (ii) interaction via local variation of the elastic modulus (modulus effect). A long-range interaction is associated with the former, and a short-range interaction with the latter. On the other hand, from the viewpoint of the distribution of solute atoms, two standpoints attaching importance to a short-range and a long-range interaction, respectively, are possible. In particular, the latter standpoint becomes important for systems with higher concentration of solute atoms, while for the dilute cases the short-range interaction is more important than the long-range interaction. Theories developed by *Fleischer* [3.19], *Friedel* [3.20] and *Suzuki* and *Ishii* [3.21] consider short-range interaction to dominate, while the theories of *Mott* and *Nabarro* [3.22], *Labusch* [3.23], and *Riddhagni* and *Asimiov* [3.24] take the former standpoint. Accordingly, these theories should be carefully applied to the appropriate range of solute concentration. Lastly, it is remarked that the elastic interaction of solute atoms with dislocations is theoretically described in the Appendix at the end of this chapter.

a) Labusch Theory. Suppose a dislocation interacts nonlocally with surrounding point obstacles. On account of this line tension, the dislocation takes a curved form, along which the maximum interaction forces F_m act sporadically. It is so because the dislocation cannot bend to interact with each solute atom, owing to its line tension, but it is affected by its average interaction force from solute atoms concentrated in the local region. Taking one of the obstacles to be the origin of the coordinates, and the number of points along the dislocation line under the influence of F to $F + dF$ per unit length of the line to be $\varrho(F)dF$, it follows under an equilibrium with the external force τb that

$$\tau b = \int \bar{\varrho}(F)\, dF = \int \varrho(y)\, F(Y)\, dy \;, \tag{3.1}$$

where an edge dislocation line is along the x-axis and the Burgers vector along the y-axis. For a solute density of c/a^2, a being the separation distance between the solvent atoms, one obtains the critical ϱ value, ϱ_c, from the condition $\partial/\partial y[\varrho(y)g(y)] = 0$. Here $g(y)$ is defined by

$$g(y) = \frac{dy(o)}{dl} \;,$$

where dl is the displacement of the dislocation at a point far from the origin. $g(y)$ is given as a function of the line tension. Inserting the critical value ϱ_c thus obtained into (3.1), $\tau_c(0)$ at $0\,\mathrm{K}$ is obtained as

$$\tau_{\rm c}(0) = \frac{c^{2/3} F_{\rm m}^{4/3} w^{1/3}}{a^{4/3}(\mu b^2/2)^{1/3} K} , \tag{3.2}$$

where μ is the rigidity of modulus, w the distance between solute atom and dislocation line giving the interaction force $F_{\rm m}$, and K the numerical factor. It is thus found that $\tau_{\rm c}(0)$ is proportional to $c^{2/3}$. Labusch's theory may be considered to be a new development of the Mott-Nabarro theory.

b) Riddhagni-Asimiov Theory. Suppose a dislocation lies in a random array of point obstacles of relatively high concentration and takes a zigzag form. *Riddhagni* and *Asimov* calculated the fluctuation of the interaction energies in this situation. If the size of the zigzag form is taken as R, the fluctuation of the interaction energy expressed by the regular distribution function is found to be proportional to $R^{1/2}$. The sum of this fluctuation energy and the energy due to the line tension, E, can be given as a function of R. Minimizing the sum with respect to R gives $R_{\rm c} \cong 200b$. Taking the average distance of unit jump as nb and writing the energy fluctuation for $R_{\rm c}$ as $\delta E_{\rm c}$ gives

$$\tau_{\rm c} b = \frac{\delta E_{\rm c}}{n b R_{\rm c}} \cong 0.1\mu \left[\left(\frac{F_{\rm m}}{\mu b^2} \right)^4 \frac{c^2(1-c)^2}{n} \right]^{1/3} . \tag{3.3}$$

At $0\,{\rm K}$, $\tau_{\rm c}$ is obtained by inserting $n = 1$, corresponding to the maximum dissipation of energy, into (3.3). $\tau_{\rm c}(0)$ is approximately proportional to $c^{2/3}$. Both the Labusch and the Riddhagni-Asimov theories yield $\tau_{\rm c}$ for $0\,{\rm K}$.[2] In the following, in order to discuss the yield stress for dillute alloys, and also at finite temperatures, the theories of *Fleischer*, *Friedel*, and *Suzuki* will be described in detail.

c) Fleischer-Friedel Theory. For dilute solutions, *Fleischer* and *Friedel* assumed a collinear arrangement between dislocation line and solute atoms under the application of external stress. Taking the average spacing between solute atoms as l_0 gives $l_0^2 = b^2/c$. In the steady-state deformation (Fig. 3.7), after the dislocation overcomes one obstacle, indicated by B, it proceeds to meet a new obstacle B'. The dislocation arc $AB'C$ is in equilibrium with the external force. When A and C are surmounted, the dislocation will again take a collinear arrangement. The

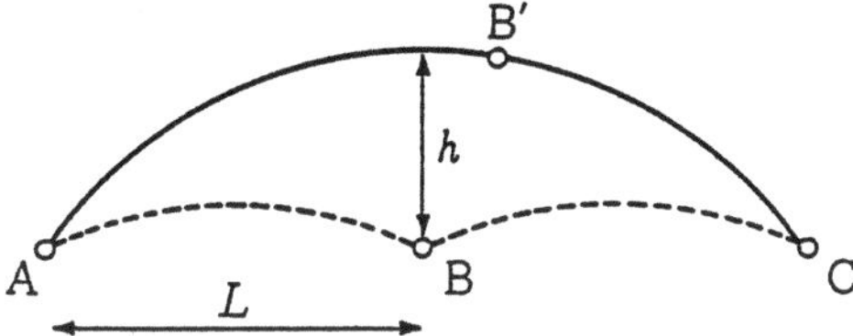

Fig. 3.7. Collinear arrangement of a dislocation with solute atoms. The dislocation meets a new obstacle B' after surmounting an obstacle B

---

[2] *Labusch* extended his theory to finite temperatures [3.25]. The final expression, however, is not given in analytical form.

steady-state motion appears by repetition of such processes. For small external stress τ satisfying the condition $h/L \ll 1$, we find

$$L^2 = 2hR \, , \tag{3.4}$$

where R is the radius of curvature of the arc $AB'C$, which gives $R = \mu b/2\tau$. The area swept out by this process is equal to l_0^2 and, therefore, $Lh = l_0^2$. Equation (3.4) is then rewritten as

$$L = \left(\frac{\mu b l_0^2}{\tau} \right)^{1/3} \, , \tag{3.5}$$

which is called *Friedel's relation*.

Taking the angle between tangents drawn at B along the arcs on the two sides as ϕ, the critical stress needed to surmount B athermally is given by the balance between the external stress and the line tension:

$$\tau_{\mathrm{c}} = \frac{\mu b}{L} \cos \frac{1}{2}\phi_{\mathrm{c}} \, . \tag{3.6}$$

Eliminating L from (3.5) and (3.6), the critical stress, designated τ_{F}, is given by

$$\tau_{\mathrm{F}} = \frac{\mu b}{l_0} \left(\cos \frac{1}{2}\phi_{\mathrm{c}} \right)^{3/2} \, . \tag{3.7}$$

τ_{F} is the yield stress at $0\,\mathrm{K}$ and is proportional to $c^{1/2}$. The factor $\cos 1/2\phi_{\mathrm{c}}$ is a measure for the strength of the obstacle:

$$\cos \frac{1}{2}\phi_{\mathrm{c}} = \frac{W}{\mu b^3} = \frac{F_{\mathrm{m}}}{\mu b^2} \, , \tag{3.8}$$

where $F_{\mathrm{m}} b = W$, F_{m} being the maximum interaction force and W the maximum interaction energy. Here, notice that *Fleischer* and *Friedel* assumed the relation $\cos 1/2\phi_{\mathrm{c}} \ll 1$, that is, their theory holds for weak obstacles, which will be clarified in Sect. 3.2.

If the dislocation surmounts B with the aid of thermal vibrations, the steady-state deformation rate is written as

$$\dot{\gamma} = nb^2 \left(\frac{l_0}{L} \right)^2 \nu_D \exp \left(-\frac{U}{k_{\mathrm{B}}T} \right) \, , \tag{3.9}$$

where ν_D is the Debye frequency and n the number of mobile dislocations. In deriving (3.9), we used relations such as

$$\left(\frac{l_0}{L} \right)^2 b\nu_D = \frac{l_0^2}{L}\nu_L = h\nu_L \, ,$$

where the vibrational frequency ν_L of dislocation of lenght L can be expressed as [3.20]

$$\nu_L = \frac{b}{L}\nu_D \ .$$

Assuming a square-shaped form for the force-distance relation between dislocation and point obstacle[3] gives

$$U = W - \tau b^2 L \quad \text{and} \quad W = F_m b \ . \tag{3.10}$$

d) Suzuki Theory. Theories worked out so far have treated the problem by replacing the random field of point obstacles with some kind of averaged field. *Fleischer* and *Friedel* assumed a collinear arrangement between dislocation line and solute atoms under an external stress. This limits accordingly the applicability of their theory to rather weak obstacles, i.e., $\cos 1/2\phi_c \ll 1$. However, *Suzuki* took a zigzag arrangement of the dislocation line through solute atoms. It will be shown that *Suzuki's* theory is applicable to a wide range of strengths of obstacle.

Let us assume the equilibrium arrangement takes a zigzag form under the application of external stress equal to τ_0, as illustrated in Fig. 3.8a, where $2xy = b^2/c$. When the external stress increases by $\Delta\tau$, i.e., $\tau = \tau_0 + \Delta\tau$, the dislocation overcomes the ith obstacle with the aid of thermal fluctuations and meets the jth obstacle. For the zigzag arrangement of the $(i-1)$th, ith, and $(i+1)$th obstacles, the interaction energy per unit length of the dislocation is

$$E_1 = \frac{W}{y} \ . \tag{3.11}$$

In Fig. 3.8b, the x-axis is taken to be parallel to the direction of the dislocation motion, and the y-axis, normal to the x-axis, is parallel to the average direction

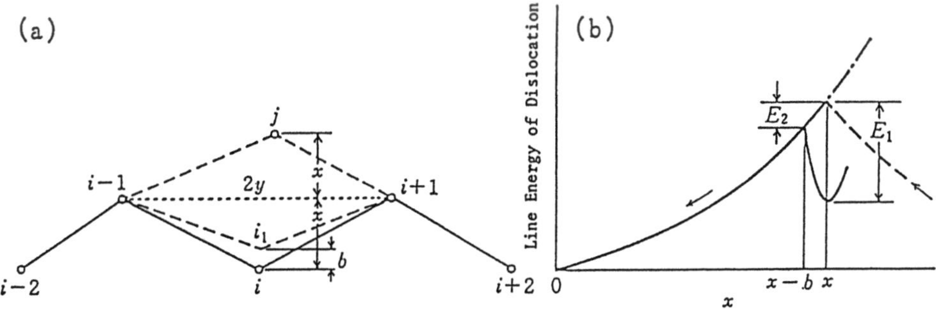

Fig. 3.8. (a) Zigzag model for dislocation motion through a random array of point obstacles. (b) Changes of energy of a dislocation line in the overcoming process from i to i_1 and j [3.5]

---

[3] If the force-distance relation is given by a second-order polynomial [3.26], the activation energy at very low temperatures is expressed as

$$U = W \left(1 - \frac{\tau}{\tau_m}\right)^{3/2} \ ,$$

where $\tau_m = F_m/bL$. In this case $\tau_c \propto c^{1/2}$ and $\tau_c \propto T^{2/3}$. See also Sect. 4.4.1.

of the dislocation line. Although this figure is drawn by assuming an attractive interaction between the dislocation and the solute atom, the following discussion will hold even for a repulsive interaction between them. The difference in the line energies at i and $i_1 = i - b$ is given for $x < y$ approximately by

$$E_2 = \frac{\mu b^3}{4y^2} (2x - b) .$$

(3.12)

The assumption $x < y$ generally holds in a dilute solution of the fcc structure, because the elastic interaction energy W is smaller than the line energy, which is equal to $\mu b^2/2$ per unit length, along the average dislocation line parallel to the y-axis.

Using the relation $2xy = b^2/c$, we can eliminate y from (3.12) and then we have

$$E_2 = \mu x^2 \left(\frac{2x}{b} - 1 \right) c^2 .$$

(3.13)

The height of the barrier is given by $|E_1 - E_2|$, as illustrated in Fig. 3.8b. From the optimum condition for the height of the barrier, i.e., $\partial |E_1 - E_2|/\partial x = 0$, x and y are determined. The result is devided into the following two regions depending upon the relative magnitude of x and b or upon the magnitude of c:

$$[1] \quad c < c_1 \quad : \quad x = \left(\frac{1}{3} \frac{W}{\mu b} \right)^{1/2} c^{-1/2} ,$$

$$y = \frac{1}{2} b^2 \left(\frac{1}{3} \frac{W}{\mu b} \right)^{-1/2} c^{-1/2} ;$$

(3.14)

$$[2] \quad c_2 > c > c_1 : \quad x = 2^{-1/3} \left(\frac{W}{\mu} \right)^{1/3} c^{-1/3}$$

$$y = 2^{-2/3} b^2 \left(\frac{W}{\mu} \right)^{-1/3} c^{-2/3} .$$

(3.15)

where c_1 is obtained by equating x and y, given by (3.14) with those given by (3.15):

$$c_1 = \frac{2^2}{3^3} \frac{W}{\mu b^3} .$$

(3.16)

While c_2 is given by putting $x = b$ in (3.15) as

$$c_2 = \frac{1}{2} \frac{W}{\mu b^3} .$$

(3.17)

Needles to say, c larger than c_2 is the region of concentrated solutions, which is beyond the scope of the present theory.

The strain rate can be written as

$$\dot{\gamma} = b^2 n \nu_D \frac{x}{y} \exp(-U/k_B T) ,$$

$$U = V - (\tau - \tau_0)b^2 y , \tag{3.18}$$

$$V = (E_1 - 2E_2)y = \frac{1}{3}W ,$$

where ν_D is the Debye frequency, n the number of mobile dislocations, and k_B the Boltzmann constant. A square-shaped force-distance relation is assumed in the above expression for U. Moreover, it should be remarked that $2E_2 y$ in V in the last equation is the sum of work done by the external stress τ_0 and the internal stress to reduce the length of the dislocation line, which is the same magnitude and acts in the same direction as τ_0. Finally, the yield stress can be obtained under the steady-state condition $\dot{\gamma}_{ext} = \dot{\gamma}$ as follows:

$$\text{For } c < c_1 , \quad \tau_c = \tau_0 + A\left(1 - \frac{T}{T_c}\right)$$

$$A = \tau_0 = \frac{\mu b}{2}\frac{x}{y^2} = \frac{2}{3^{3/2}}\frac{W^{3/2}}{\mu^{1/2}b^{9/2}}c^{1/2} \tag{3.19}$$

$$T_c = \frac{1}{3}W \bigg/ \left[k_B \ln\left(\frac{2}{3}\frac{W}{\mu}\frac{n\nu_D}{b\dot{\gamma}_{ext}}\right)\right] .$$

$$\text{For } c_2 > c > c_1 , \quad \tau_c = \tau_0 + B\left(1 - \frac{T}{T_c}\right) ,$$

$$B = \tau_0 = \frac{\mu b}{2}\frac{x}{y^2} = \frac{W}{b^3}c , \tag{3.20}$$

$$T_c = \frac{1}{2}W \bigg/ \left[k_B \ln\left(\frac{W^{3/2}}{\mu^{3/2}}\frac{2cn\nu_D}{\dot{\gamma}_{ext}}\right)\right] .$$

According to the above equations, two statistical regions exist, for $c < c_1$ and $c_2 > c > c_1$, respectively. For $T \geq T_c$, $\tau_c = \tau_0$, which forms the plateau region. The concentration dependence of τ_0 is the same as τ_c in both regions. Furthermore, it is noticed that τ_c at $T = 0$ is given by $2\tau_0$. Writing τ_c at $T = 0$ (3.19) as τ_S, we find that

$$\tau_S = 0.78\tau_F , \tag{3.21}$$

where τ_F is τ_c at $T = 0$ given by *Friedel*, eq. (3.7).

Figure 3.9 illustrates the relation between the strength of a point obstacle expressed in terms of $W/\mu b^3 = \cos(\phi_c/2)$ and $\tau_c(0)$ as a function of c. It shows three statistical regions, in which the third region, $c > c_2$, is the one discussed by *Mott-Nabarro, Labusch,* and *Riddhagni-Asimov* to give $\tau_c \propto c^{2/3}$. The first and second regions have already been treated in (3.10, 19, 20). Recently, *Schwarz* and *Labusch* [3.27] concluded the existence of the second region by means of an approach different from *Suzuki*'s. Meanwhile, *Kocks* et al. [3.28] have derived independently a diagram similar to Fig. 3.3.

In Cu-Ni alloys, the elastic interaction energy W is calculated, using the method described in the Appendix to this chapter, to be 0.11 eV. Inserting this

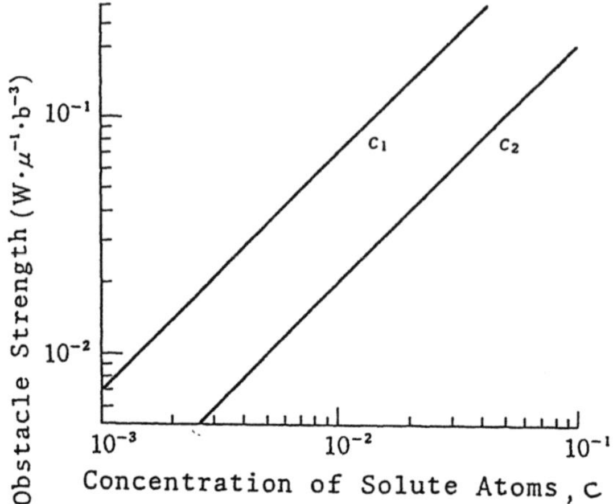

Fig. 3.9. Strength vs concentration of point obstacles, indicating three statistical regions with respect to concentration dependence of yield stress (see the text) [3.30]

value into (3.16), we have $c_1 = 2.5 \times 10^{-2}$. We obtained $c_1 = 2 \times 10^{-2}$ experimentally, as shown in Fig. 3.2, which agrees very well with the above calculation.

3.2 Comparison of Theories of Solution Hardening with Computer Simulation

It is of interest to compare the theories given by *Fleischer* and *Friedel* and by *Suzuki* with the result of the computer simulation carried out by *Foreman* and *Makin* [3.29]. Figure 3.10 shows the relation of τ_c at 0 K to the strength of point obstacles ϕ_c [3.30]. $\phi_c = 0$ is for a strong obstacle, where the dislocation overcomes the obstacle by the Orowan mechanism [3.31], while $\phi_c = \pi$ is for zero strength. The three results almost coincide near $\phi_c = \pi$. In this limit, for instance, the *Suzuki* model perfectly coincides with the model of *Fleischer* and *Friedel*. The difference between them becomes clear with increasing strength of obstacle. The computer simulation used the so-called rolling method, which seems to simulate well the behavior of dislocations moving through a random distribution of point obstacles. As seen from Fig. 3.10, the latter deviates sharply from the Fleischer-Friedel theory with decreasing ϕ_c, while it is closer to Suzuki's theory in the whole range of ϕ_c. It is concluded, therefore, that the zigzag model is better than the collinear model at representing the average behavior of dislocations moving through a random distribution of point obstacles. The computer simulation gave $\tau_c = 0.8\tau_F$, which agrees with τ_S, since according to (3.21) τ_S is equal to $0.78\tau_F$.

The three expressions for the activation energy U_0 can be given in terms of x and y for the zigzag model as follows:

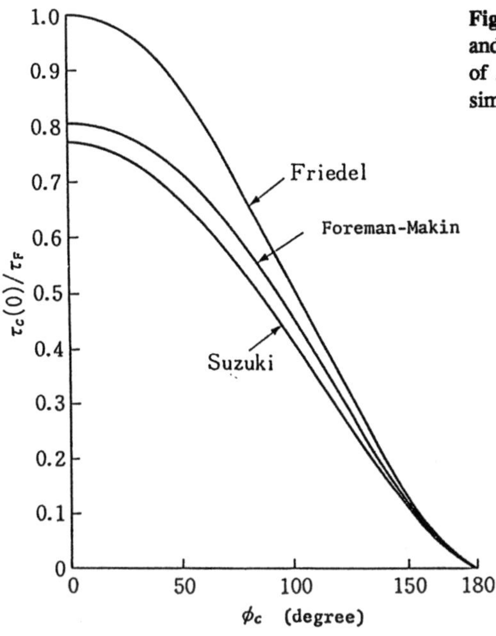

Fig. 3.10. Relations between yield stress at 0 K and obstacle strength ϕ_c obtained by the theories of *Friedel* and of *Suzuki* and by the computer simulation by *Foreman* and *Makin* [3.30]

Friedel : $\qquad U_0 = \mu b^3 \cos(\phi_c/2) = W$,

Foreman-Makin : $\qquad U_0 = \mu b^3 [\cos(\phi_c/2) - (x/y)]$,

Suzuki : $\qquad U_0 = \mu b^3 [\cos(\phi_c/2) - (x/2y)] = \tfrac{2}{3} W$.

In the course of the rolling procedure used by the computer simulation, the dislocation often takes a zigzag form locally before surmounting one of randomly distributed obstacles, which reduces U_0 from W by $\mu b^3 x/y$ on account of the geometric effect. Finally it is also remarked that the collinear model yields the largest U_0, and accordingly it tends to overestimate τ_c in comparison with the other two approaches.

3.3 Effect of a Random Distribution of Point Obstacles on τ_c

Foreman and *Makin* [3.29] found by computer simulation an unzipping of dislocations moving through a random distribution of point obstacles. An obstacle separated by a greater distance from neighboring obstacles is first overcome, inducing a sequence of unzipping through the array of obstacles along the dislocation line. The phenomenon is shown in Fig. 3.11, where l is the separation distance of point obstacles given by $(b^2/c)^{1/2}$. Due to the static aspect of the simulation, however, this unzipping does not have much influence upon the es-

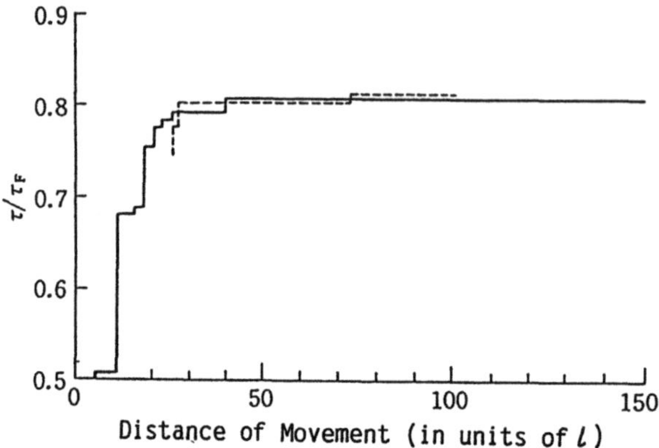

Fig. 3.11. Unzipping effect appearing in computer simulation [3.29]

timation of yield stress. In fact, *Foreman* and *Makin* obtained $\tau_c(0) = 0.8\tau_F$, which is equal to τ_S as mentioned before. In mechanical testing in the laboratory, however, measurements of yield stress are carried out at a constant rate of deformation. Accordingly, the unzipping effect, i.e., the effect due to deviation from the average distribution of point obstacles, has a great influence on experimental results. Not only the static process of surmounting under an applied stress but also the process with the aid of thermal activation will be greatly influenced, so that the strain rate may become much larger than the case in which unzipping does not occur. According to *Suzuki* [3.4], $\tau_S = q\tau_y(0)$, where $q = 5.1$ for Cu-Ni alloys. $\tau_y(0)$ is given by the extrapolation of yield stresses obtained below room temperature to $0\,\mathrm{K}$, where the anomaly that occurs at temperatures below $100\,\mathrm{K}$ (Fig. 3.5) is neglected.

If we assume q not to be so sensitive to temperature and insert $q\tau$ in place of τ in (3.18), the steady-state velocity of dislocations is given as

$$v = \frac{x}{y}\nu_D b \exp\left(-\frac{(2/3)W - q\tau b^2 y}{k_B T}\right) . \tag{3.22}$$

The solid lines for the Cu-Ni alloy at $300\,\mathrm{K}$ and $77\,\mathrm{K}$ in Fig. 2.7a are calculated from (3.22) using the values $q = 5.1$ and $W = 0.11\,\mathrm{eV}$. The calculated curves shown in Fig. 3.5 are $\tau_c(T)$ curves obtained using the above values for q and W. They are almost in agreement with the observed values.

As described above, the effect of unzipping caused by the random distribution of point obstacles can well be taken into account by introducing the factor q. It expresses an increase of the separation distances between obstacles from (x, y) to (qx, qy); in other words, the number of effective obstacles is reduced by the factor $1/q^2$. A statistical treatment of q will be given in the next chapter, where the W dependence of q will be described.

3.4 Appendix: Elastic Interaction Between a Dislocation and a Solute Atom

We will summarize the studies by *Fleischer* [3.32], *Nabarro* [3.33], *Seeger* and *Haasen* [3.34], and *Sax* [3.35]. An electrostatic interaction is small compared to the elastic interaction and so it is not discussed here. A chemical interaction is also excluded from the discussion for the reason discussed in Sect. 3.1.2.

(1) Size effect: A solute atom producing an isotropic dilation field interacts with a hydrostatic field around an edge dislocation. If we write Poisson's ratio as ν, the shear modulus as μ, and the volume of a solute atom as Ω, the interaction energy of the solute atom at (r, θ) from the center of a positive dislocation is given in polar coordinates by

$$W_1^e = \frac{3\mu b \Omega \sin \theta}{\pi r} \varepsilon_a \,, \tag{3.23}$$

where $\varepsilon_a = d \ln a / dc$, c is the concentration of the solute atoms and a the lattice constant. The sign of an edge dislocation is defined as positive or negative according to whether the extra half-plane lies above or below the slip plane.

For a screw dislocation, a solute atom interacts with the nonlinear stress field in the core region of the dislocation, and the interaction energy is expressed by

$$W_1^s = \frac{3\lambda}{2\pi^2} \frac{1-\nu}{(1-2\nu)} \mu\Omega \frac{b^2}{r^2} \varepsilon_a \,, \tag{3.24}$$

where $\lambda \sim 0.25$ [3.36].

(2) Modulus effect: A solute atom changes the shear modulus and the bulk modulus of its surrounding matrix. The interaction energy caused by changes of elastic moduli is given by

$$W_2^e = \frac{\mu b^2 \Omega \varepsilon_\mu}{8\pi^2(1-\nu)^2 r^2}(1 - \varepsilon_{\mu K} \sin^2 \theta) \,, \tag{3.25}$$

where, taking $\nu = 1/3$, it holds that

$$\varepsilon_{\mu K} = 1 - 0.3(\varepsilon_K / \varepsilon_\mu) \,,$$

$$\varepsilon_K = d \ln K / dc \,, \quad \varepsilon_\mu = d \ln \mu / dc \,.$$

Meanwhile the total interaction energy $(W_1^S + W_2^S)$ for a screw dislocation is given by the sum of W_1^S and W_2^S where $\varepsilon_K = 0$ is taken. *Takeuchi* [3.37] calculated the change of the interaction energy (W_2^e) due to the change of the elastic modulus of the matrix introduced by the long-range interaction with the solute atom of an edge dislocation. The magnitude of the change, however, was found to be smaller than W_2^e given by (3.25).

Table 3.1 illustrates the results of calculations of elastic interaction energies for solute atoms in copper given by *Saxl*, in which the values of W_1^e have been recalculated by *Nabarro* [3.36].

Table 3.1. The maximum elastic interaction energy between a dislocation and a solute atom in copper [3.35]

	W_1^e [eV]	W_2^e [eV]	W_1^s [eV]	W_2^s [eV]
Al	0.09	0.013	0.060	0.014
Si	0.028	0.026	0.018	0.019
Zn	0.08	0.023	0.051	0.021
Vacancy	0.028	0.16	0.018	0.072

4. Dislocation Dynamics and Strength of Crystalline Materials

The experimental discovery of a remarkable phenomenon of loss of strength in superconducting metals and alloys will first be described, followed by a similar phenomenon due to the rapid decrease of the number of phonons at low temperatures. This loss of strength of metallic materials definitely surpasses all imagination. In this chapter, these phenomena are discussed in detail. Finally dynamic theories of the strength of materials are described taking account of the interaction of moving dislocations with electrons and phonons (inertial effects) as well as the effect of random distributions of obstacles on the dislocation motion (unzipping effect). We conceive that these studies may contribute to the development of research on the strength of materials from the qualitative to the quantitative level.

4.1 The Loss of Strength of Metals and Alloys in the Superconducting State

The loss of strength in the superconducting state was first discovered in lead and niobium by *Kojima* and *Suzuki* [4.1]. Since then, *Pustovalov* et al. [4.2] and many others have studied this phenomenon in various metals and alloys of fcc, hcp, and bcc structures. Excellent review articles on this problem have been published [4.3–8].

The discovery of this phenomenon was dramatic in that it instantly disclosed the basic importance of dislocation dynamics in understanding the strength of metals and alloys. It awakened our interest in a similar but new phenomenon: the loss of strength in the normal-conducting state of metals and alloys at low temperatures, which was first found in Cu-Ni alloys and was mentioned in Chap. 3. The decrease of electron frictional force against the motion of dislocations causes the former phenomenon in the superconducting state, and the decrease of the phonon frictional force causes the phenomenon in the normal state. A new theory on the inertial effect of moving dislocations in relation to the first phenomenon was worked out by *Granato* [4.9] and *Suenaga* and *Galligan* [4.10].

Figure 4.1 illustrates the stress-strain curve of polycrystalline lead at 4.2 K [4.1]. The letters s and n indicate the superconducting and normal states, respectively. During deformation in the s state, when an external magnetic field larger than the critical value (H_c) is applied (ON), the specimen is transformed to the n

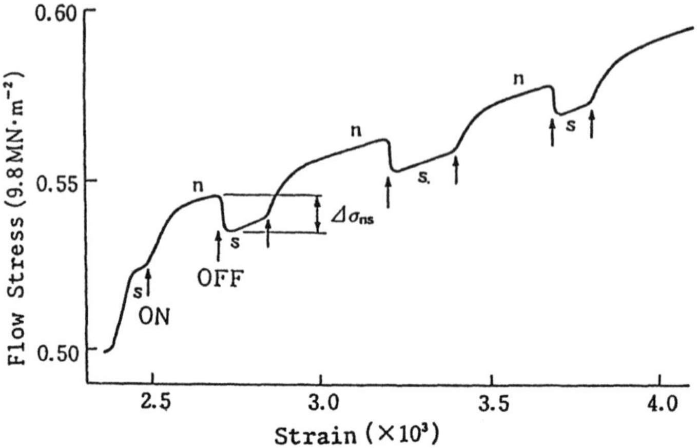

Fig. 4.1. Stress-strain curve of polycrystalline lead at 4.2 K measured at a strain rate of $4 \times 10^{-5}\,\mathrm{s}^{-1}$. The specimen is in the superconducting state (s) under zero magnetic field (OFF), and is transformed into the normal state (n) by the application of a magnetic field larger than H_c (ON) [4.1]

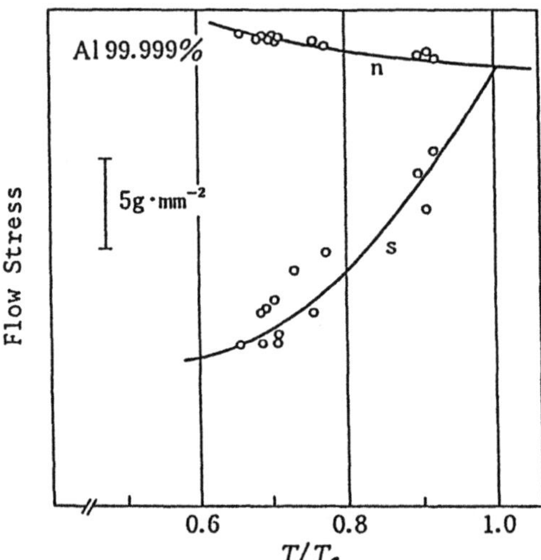

Fig. 4.2. Temperature dependence of flow stress at the early stage of deformation in the superconducting (s) and normal (n) states in aluminum [4.12]. The superconducting phase transition temperature is equal to 1.18 K and the strain rate used is $6.9 \times 10^{-5}\,\mathrm{s}^{-1}$

state and the flow stress increases by $\Delta\sigma_{ns}$. When the magnetic field is removed (OFF), the flow stress decreases again to the extension of the former level at the s state. The variation of the flow stress with magnetic field is quite reversible. Although the following illustrations were all obtained on single-crystalline materials, the loss of strength in the superconducting state is a universal phenomenon in poly- and single-crystalline materials and also independent of crystal structure. Figure 4.2 shows the yield stress of aluminum as a function of temperature in the s and n states. The transition temperature from the normal to the superconducting state, T_e, is equal to 1.18 K [4.11, 12].

In the following sections, important features of $\Delta\tau_{ns}$ disclosed by many experiments are described.

4.1.1 Temperature Dependence

Values of $\Delta\tau_{ns}$ obtained from Fig. 4.2 are plotted against the reduced temperature T/T_e in Fig. 4.3. Form this figure, $\Delta\tau_{ns}$ is found to be proportional to $(1 - \Gamma_{BCS})$, where Γ_{BCS} is given by

$$\Gamma_{BCS} = \frac{2}{1 + \exp[\Delta(T)/k_B T]} \tag{4.1}$$

where $\Delta(T)$ is the energy-gap function. According to *Bardeen*, *Cooper*, and *Schrieffer* [4.13], Γ_{BCS} is equal to the ratio of acoustic attenuations in the two states α_s/α_n, and so it can be written analogously as

$$\Gamma_{BCS} = B_s/B_n , \tag{4.2}$$

where B_s and B_n are the frictional coefficients associated with the motion of dislocations due to the scattering of conduction electrons in the two states (Chap. 3). As discussed theoretically by *Brailsford* [4.14] and others, B_n is independent of temperature. On the other hand, B_s decreases to zero at $T = 0$ as the number of conduction electrons decreases, and so the inertial effect increases, causing a loss of strength in the s state, which will be theoretically described in Sect. 4.3. The proportionality of $\Delta\tau_{ns}$ to $(1 - \Gamma_{BCS})$ has been established in many metals and alloys.

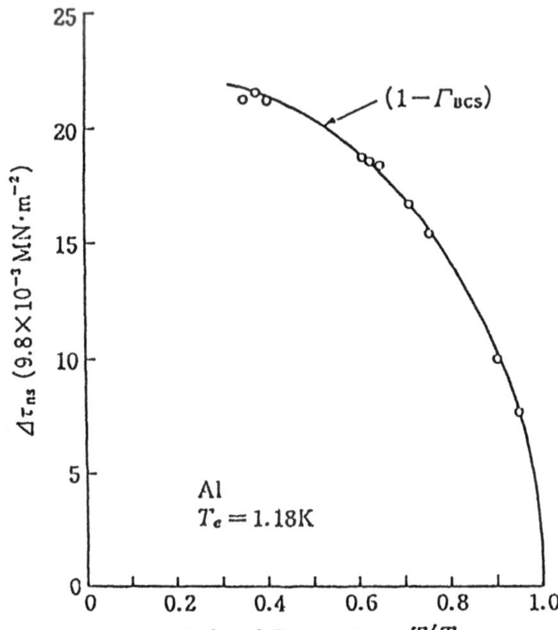

Fig. 4.3. $\Delta\tau_{ns}$ of aluminum as a function of reduced temperature

4.1.2 Impurity Dependence

In the case of alloys, $\Delta\tau_{ns}/\tau_n$ generally increases with solute concentration c and it reaches as much as 20% in some alloys [4.15]. According to *Kostorz* [4.15],

$$\Delta\tau_{ns} = \Delta\tau_{ns}^0 + Kc^{1/2} \, , \tag{4.3}$$

where $\Delta\tau_{ns}^0$ is $\Delta\tau_{ns}$ for pure metals, and K is a proportionality constant. The above relation suggests that $\Delta\tau_{ns}$ is closely related to solution hardening caused by the elastic interaction of moving dislocations with solute atoms as discussed in Chap. 3. In fact, *Suenaga* and *Galligan* [4.10] found that $\Delta\tau_{ns}$ depends linearly on $\Delta a/a$, the rate of variation of the lattice parameter per unit concentration of solute atoms, but not on the rate of variation of the electron density.

4.1.3 Strain-Rate Dependence

It is important to know whether $\Delta\tau_{ns}$ is caused by the athermal or thermal process of dislocations surmounting point obstacles. *Alers* et al. [4.16] first tried to clarify this problem. As already explained in Chap. 3, the flow stress of crystals is determined by the distribution and strength of obstacles. There are two cases: mobile dislocations overcome obstacles either by the athermal process or by the thermal activation process. For the latter case, the activation process is given by (3.10) as

$$U = W - \tau_e V \, , \tag{4.4}$$

where $\tau_e = \tau - \tau_i$, (2.4), and V is the activation volume. Increasing the strain rate from $\dot{\gamma}_1$ to $\dot{\gamma}_2$, and assuming no change in the density of mobile dislocations, we find (3.9) gives the increment of the flow stress as

$$\Delta\tau = \frac{k_B T}{V} \ln(\dot{\gamma}_2/\dot{\gamma}_1) \, . \tag{4.5}$$

Since the temperature is very low, $\Delta\tau$ for $\dot{\gamma}_2/\dot{\gamma}_1 = 10$, for example, will be very small. On the other hand, for the athermal process (2.2) and (2.8) give

$$\dot{\gamma} = bnv = b^2 n\tau/B \, . \tag{4.6}$$

Accordingly, if $\dot{\gamma}_{ext}$ increases by a factor of 10, and ignoring changes in other quantities in (4.6), the flow stress becomes ten times larger.

Figure 4.4, taken from [4.17], illustrates the effect of strain rate upon the flow stress of lead below T_e. As seen from this figure, the strain-rate sensitivity of flow stress is very small and different in the s state from the n state, the latter being slightly larger than the former. This is a sure indication that the thermal activation process is favored compared to the athermal process. As will be discussed in Sect. 4.4.4, the activation volume V in the n state is smaller than in the s state, and accordingly, $\Delta\tau$ given by (4.5) is larger in the former than the latter. The above strain-rate sensitivity is called the normal sensitivity.

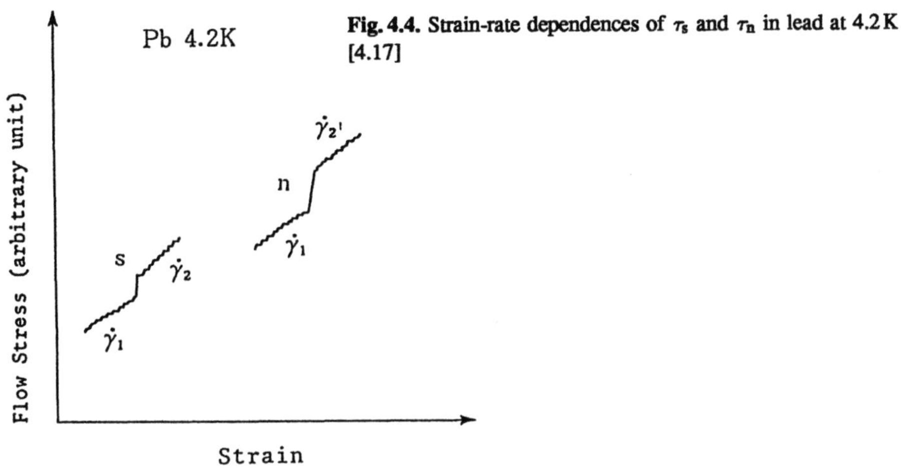

Fig. 4.4. Strain-rate dependences of τ_s and τ_n in lead at 4.2 K [4.17]

4.1.4 Anomalous Strain-Rate Dependence

Figure 4.5 shows the $\Delta\tau_{ns}$ results obtained for Al-Mg 0.3 at.% alloy (T_e = 1.15 K), where filled circles correspond to $\Delta\tau_{ns}$ for $\dot{\gamma} = 4.6 \times 10^{-5}$ s^{-1} and open circles are for $\dot{\gamma}_2 = 4.6 \times 10^{-4}$ s^{-1}. The strain-rate sensitivity of $\Delta\tau_{ns}$ is quite large, contrary to the case discussed in Sect. 4.1.3. When the alloy is deformed at $\dot{\gamma}_2$, the specimen temperature increases by as much as 0.2 K, so it is necessary

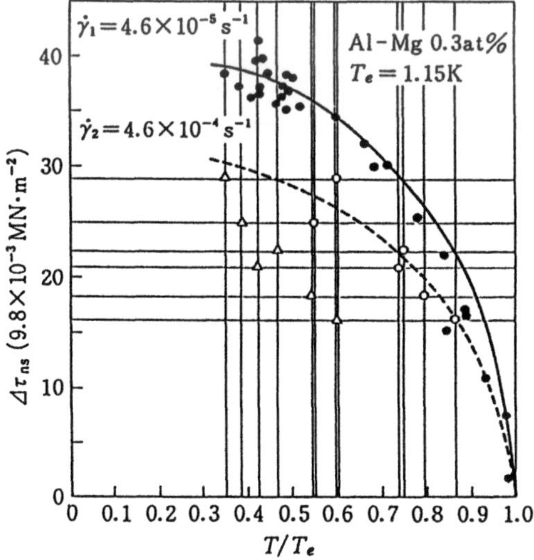

Fig. 4.5. Strain-rate dependence of $\Delta\tau_{ns}$ as a function of temperature for Al-Mg 0.3 at.% alloy. The specimen temperatures have been corrected for $\dot{\gamma} = 6 \times 10^{-4}$ s^{-1} by the following method: H_c of the specimen was measured during deformation at each temperature and compared with the H_c-T curve measured on the same specimen under no external stress. (o): after correction; ($\triangle$): before correction [4.11]

to measure it exactly during deformation. By fitting values of the critical magnetic field H_c measured during deformation to the H_c-T curve obtained from the specimen under no external stress, we can determine the specimen temperature under deformation. In Fig. 4.5, triangles indicate $\Delta\tau_{ns}$ before the correction and open circles give the results after correction of the specimen temperature.

The above strain-rate dependence of $\Delta\tau_{ns}$ is not due to that of τ_s shown in Fig. 4.4, which is negligibly small at $T < 1.15\,\mathrm{K}$, but instead to the anomalous strain-rate sensitivity of B_s. As will be theoretically explained in Sect. 4.3, B_s is expected to increase when the dislocation velocity exceeds a certain critical velocity. Because the critical velocity is given by a linear function of the energy-gap function $\Delta_0(0)$, the velocity of dislocations in aluminum and aluminum alloys, the transition temperatures of which are around $1.15\,\mathrm{K}$, is expected to easily overcome the critical velocity.

4.1.5 Strain Dependence

The variation of $\Delta\tau_{ns}$ with shear strain γ is not simple. It increases with strain in lead and indium [4.18–21], while it decreases with strain in aluminum and aluminum alloys [4.11]. In the latter, the strain dependence of $\Delta\tau_{ns}$ in the easy glide region differs from that in the second hardening stage of deformation. This behavior of $\Delta\tau_{ns}$ is thought to be caused by the anomalous strain-rate sensitivity of B_s as mentioned before.

4.2 Loss of Strength in the Normal State of Solid Solutions at Low Temperatures

The phenomenon of the loss of strength below $100\,\mathrm{K}$ of normal metals and alloys was first found in Cu-Ni alloys (Fig. 3.5) [4.22]. Since then a number of studies have been carried out on various alloys, e.g. Cu-Ag [4.23], Cu-Al [4.8], Cu-Ge [4.24], Cu-Co [4.25], Ag-In [4.26], Ag-Sn [4.26], and Al-Mg [4.11]. The result obtained on Cu-Al alloys is shown in Fig. 4.6 [4.8]. As seen from this figure, the yield stress begins to decrease from the critical temperature T_p and increases again near $0\,\mathrm{K}$. Peaking of yield stress appears in solid solutions of relatively low concentration. *Schwarz* et al. [4.27] studied this phenomenon by the method of internal friction. Their results on Cu, Cu-Al alloy, Pb, and Pb-Sn alloy are shown in Fig. 4.7, where the stress amplitudes giving a constant decrement Δ are plotted as a function of temperature. Since the stress amplitude is the stress necessary for the vibrational motion of dislocations inducing a constant decrement, it can be assumed to have some close relation with the yield stress.

Peak temperatures of Cu-Al alloys are about $50\,\mathrm{K}$ (Fig. 4.6), which coincide with those equal to about $52\,\mathrm{K}$ illustrated in Fig. 4.7. The concentration dependence of the peak temperature T_p is shown in Fig. 4.8 for Cu-Al [4.8] and Ag-In [4.26] alloys, from which we conclude that T_p is proportional to $c^{1/2}$.

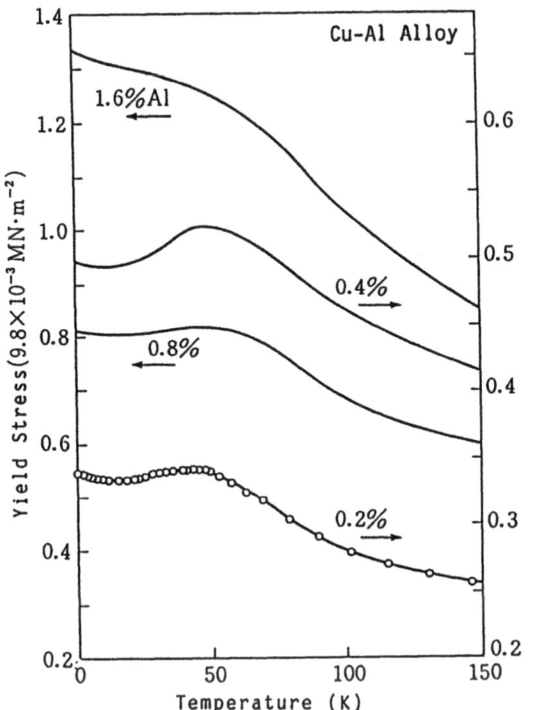

Fig. 4.6. Temperature dependence of τ_c measured on Cu-Al alloys. The atomic concentration of aluminum is given by each curve. The peaks of $\tau_c(T)$ appear around 50 K [4.8]

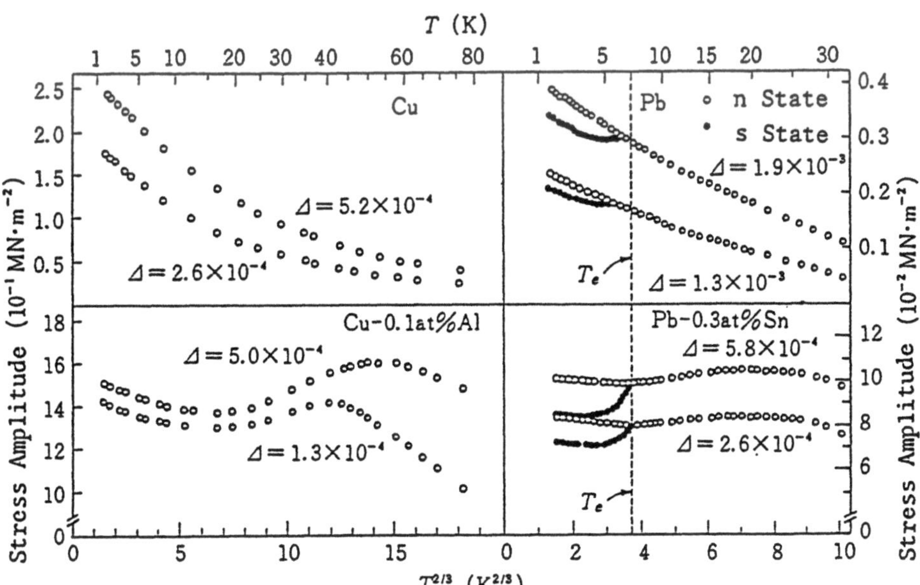

Fig. 4.7. Stress amplitude corresponding to a constant decrement Δ plotted against temperature for Cu, Cu-Al, Pb and Pb-Sn [4.27]

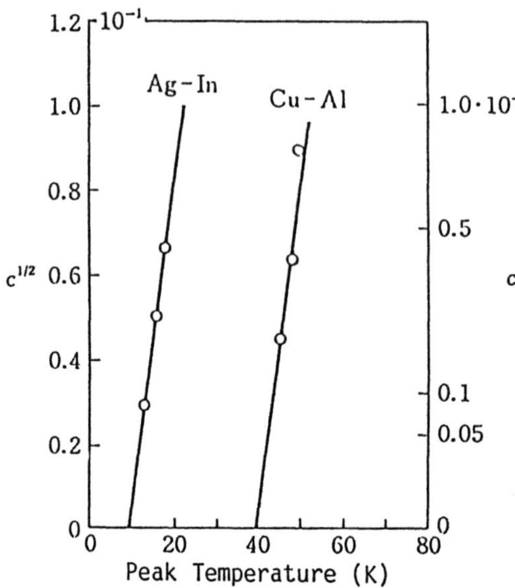

Fig. 4.8. Measured T_p plotted against c for Ag-In [4.26] and Cu-Al [4.8] alloys. The peak temperature T_p is shown to be proportional to $c^{1/2}$

4.3 Theory of Inertial Effects

4.3.1 Inertial Theory

Granato [4.9] and *Suenaga* and *Galligan* [4.10] independently developed the so-called inertial theory of dislocation motion in the field of point obstacles in order to explain the loss of strength induced by the transition form the normal to the superconducting state.

Consider a dislocation approaching the line $\overline{OX}$ under an external stress τ, where point obstacles sit at O and X, as illustrated in Fig. 4.9a. A damped oscillation of the dislocation line element $\overline{OX}$ occurs if the frictional coefficient B satisfies the following condition (underdamped condition):

$$\beta = \frac{B}{2A} < \omega_0 , \tag{4.7}$$

where ω_0 is the fundamental frequency of the dislocation line of length L and A is the effective mass of the dislocation. Writing the density of the material as ϱ gives $A = \varrho b^2$. Using the line tension C, ω_0 is given in (2.12) as

$$\omega_0 = \frac{\pi}{L}\left(\frac{C}{A}\right)^{1/2} , \tag{4.8}$$

where $C = \mu b^2/2$. The damped oscillation as a function of time t is illustrated in Fig. 4.9b, in which the displacement of the dislocation line at the midpoint of the arc is expressed by ξ. If we write the maximum displacement as ξ_{max}, the

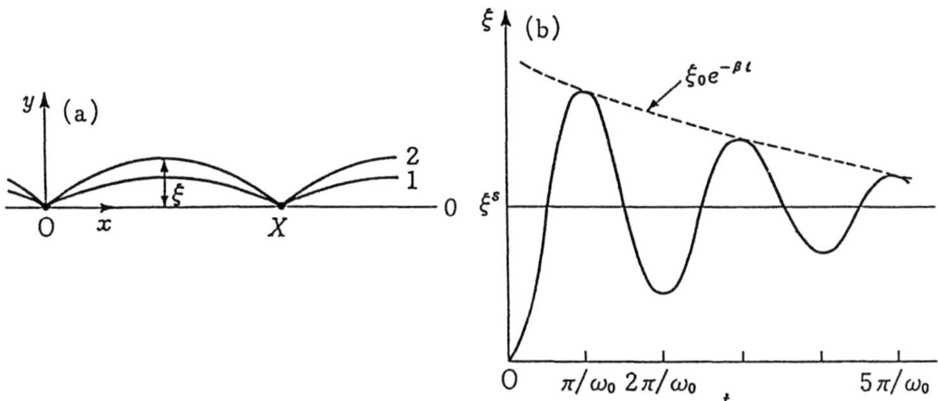

Fig. 4.9. (a) Schematic illustration of the inertial effect. **(b)** Underdamped oscillation of a dislocation segment pinned by obstacles at O and X under an external stress as a function of time

y component of the line tension exerts the maximum force in the direction of the dislocation motion on the obstacles at O or X when $\xi = \xi_{\max}$; the maximum force is given by

$$F_{\max} = b\tau L Y_{\max} ,$$
$$\text{where} \quad Y_{\max} = 1 + \mathrm{e}^{-z} \quad \text{and} \quad z = \pi\beta/\omega_0 . \tag{4.9}$$

The force acting on the dislocation is a factor $Y_{\max}$ larger than the one for static displacement ξ^s, determined by the equilibrium between the external force $b\tau L$ and the line tension C.

Since the same magnitude of force is needed for the dislocation to overcome the obstacle irrespective of the n or s state, $\Delta\tau_{\mathrm{ns}}(0)$ at $T = 0\,\mathrm{K}$ is obtained from the equation $(F_{\max})_s = (F_{\max})_n$ as

$$\Delta\tau_{\mathrm{ns}}(0) = \frac{L}{4(CA)^{1/2}}(B_{\mathrm{n}} - B_{\mathrm{s}})\tau_{\mathrm{n}} .$$

Taking $B_{\mathrm{s}}(0) = 0$, we have

$$\Delta\tau_{\mathrm{ns}}(0) = \frac{L}{4(CA)^{1/2}}B_{\mathrm{n}}\tau_{\mathrm{n}} . \tag{4.10}$$

For further development of the above theory, we have to note the following two points. (1) The above theory is only valid at $T = 0\,\mathrm{K}$. We have to extend it, therefore, to a theory applicable at finite temperatures. This is because the thermal activation process is based on probability theory and, accordingly, we cannot describe it by using the probability of overcoming a barrier at a definite time such as $t = \pi/\omega_0$. (2) $Y_{\max}$ given by (4.9) enhances the deformation of the substance similarly to but independently of the factor q for the motion of dislocations in the random arrays of obstacles given by (3.22).

Meanwhile, we have only been concerned with the electron frictional coefficient B_e so far in the above theory. Similar discussion is possible for the phonon frictional coefficient B_p, which is able to explain the loss of strength observed in solid solutions below T_p.

Lastly, we mention the point raised by *Natsik* [4.28] that not only the displacement ξ but also the vibrational frequency ν_L, appeared in (3.9) and (3.18), may depend upon B. According to *Granato* [4.29], however, ν_L does not depend upon B as long as $B \ll 2A\omega_c$, where ω_c is the angular frequency of the dislocation at the top of the potential due to an obstacle.

4.3.2 Excitation of Quasiparticles by Moving Dislocations and Anomalous Strain-Rate Sensitivity of B_s

When a dislocation moves at high speed, two excited electrons (quasiparticles) may be formed by the collision with a Cooper pair in the ground state. This process should be accompanied by loss of kinetic energy of the dislocation to induce a new frictional term. The critical velocity of a dislocation for the destruction of a Cooper pair is given as [4.30, 31]

$$v_c = \frac{2\Delta_0}{\hbar q_D} \, , \tag{4.11}$$

where $\hbar$ is Planck's constant divided by 2π, Δ_0 is the energy-gap function at 0 K and q_D is the Debye wave number of the moving dislocation. For aluminum, v_c

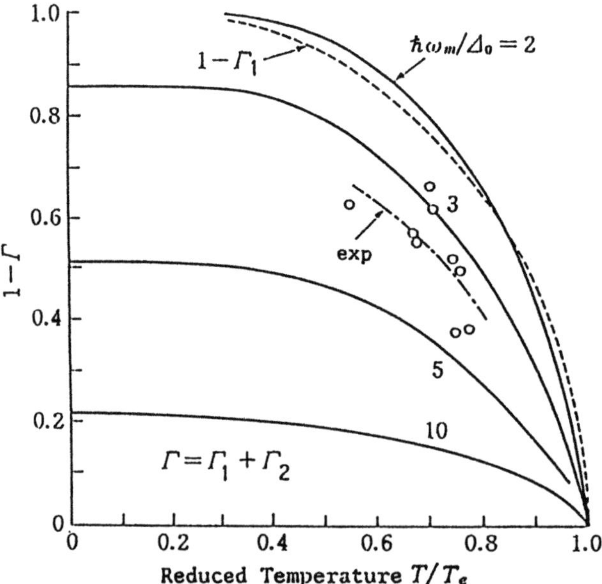

Fig. 4.10. Dependence of $(B_e)_s$ on the dislocation velocity [4.30]. Chained curve and open circles are obtained from the data for aluminum deformed at the strain rate $\dot{\gamma}_2$ shown in Fig. 4.5

is estimated to be 3×10^3 cm/s, which is the order of magnitude to be expected for the dislocation at the yield point at low temperatures (Table 2.1). Destruction of Cooper pairs may occur in aluminum with low T_e, i.e., small Δ_0, while the dislocation velocity hardly exceeds the critical velocity V_c in lead, which has a higher T_e, i.e. large Δ_0. The anomalous strain-rate sensitivity found in aluminum alloys, as discussed in Sect. 4.1.4, but not in lead and lead alloys can be explained by the present mechanism.

According to the above theory,

$$\Gamma = \Gamma_1 + \Gamma_2 \ , \tag{4.12}$$

where $\Gamma_1 = \Gamma_{\text{BCS}}$.

Γ_{BCS} is given by (4.1) and Γ_2 is the additional term due to the formation of quasiparticles. $1 - \Gamma(v)$ is plotted against T/T_e in Fig. 4.10 [4.11]. The experimental curve for $\dot{\gamma}_1 = 4.6 \times 10^{-5}\,\text{s}^{-1}$ in Fig. 4.5 (Al-Mg 0.3 at.% alloy) is fitted to the curve $\hbar\omega_{\text{m}}/\Delta_0 = 2$ and experimental points for $\dot{\gamma}_2 = 4.6 \times 10^{-4}\,\text{s}^{-1}$ are plotted in this figure. From this, we can estimate $\hbar\omega_{\text{m}}/\Delta_0 = 3.5$ for $\dot{\gamma}_2$, that is, $v = \omega_{\text{m}}/q_{\text{D}} = 3.5\Delta_0/\hbar q_{\text{D}} = 5.6 \times 10^3$ cm/s, which is definitely larger than $v_{\text{c}} = 3 \times 10^3$ cm/s.

4.4 Quantitative Treatment of the Strength of Metals and Alloys of fcc Structure

The foregoing theories of the strength of metals and alloys of fcc structure have been quasistatic in nature and used an average field approximation for the random arrays of obstacles to the motion of dislocations. As a result, theory overestimates τ_c by a factor of 5 compared to experiments on Cu-Ni alloys. To raise the present state of theories from the qualitative to the quantitative level, we have to take account of the dynamical properties of moving dislocations as well as the effect of deviation from the average distribution of obstacles assumed in them. The first point to be noted is the q factor discussed in the derivation of (3.22), which is due to the unzipping effect of moving dislocations through a random array of point obstacles. For the quantitative estimation of the strength of metals and alloys, it is now clear that we cannot use the simple average distribution of point obstacles. The next point to be noted is the effects of inertia due to electron and phonon frictional forces on the process of surmounting point obstacles, which also produce drastic changes in the strength of metals and alloys at low temperatures.

4.4.1 Unzipping Effect

Drawing parallel lines through the minima of Peierls potential for the primary edge dislocation in the slip plane in the lattice of a solid-solution alloy, we can write the distribution function of point obstacles along the kth line of unit length as

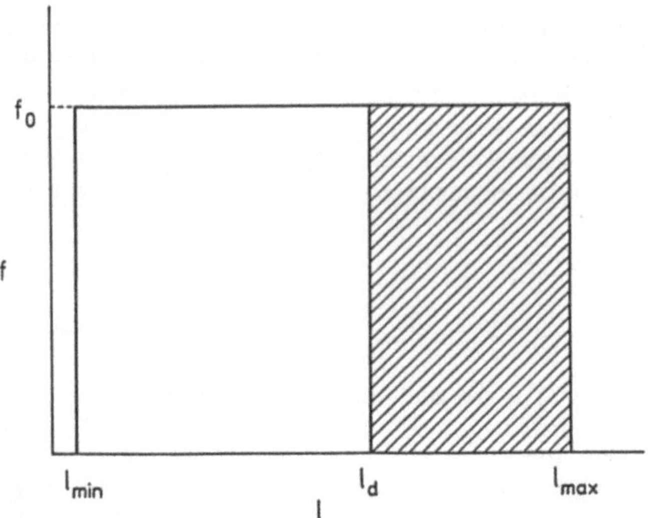

Fig. 4.11. Distribution of point obstacles along a dislocation line under zero stress as a function of the characteristic length of a point obstacle defined by half the sum of the distances between neighboring obstacles. Obstacles with $l \geq l_d$ are overcome under an external stress τ without the aid of thermal activation

$$f_k = f_k(l_i) , \tag{4.13}$$

where l_i is the characteristic length of the ith obstacle equal to half the sum of the distances to the nearest obstacles along the line. For the sake of brevity, we take a simple distribution function independent of k as shown in Fig. 4.11, i.e.,

$$f_k = f_0 = \text{const} ,$$

and $l_{\min} = b$ is taken to be negligible compared to $l_{\max}$. Since the number of obstacles per unit area is given by

$$(1/b) f_0 l_{\max} = c/b^2 ,$$

we have

$$f_0 = c/b l_{\max} . \tag{4.14}$$

If we suppose that obstacles with characteristic lengths $l_i \geq l_d$ are all overcome athermally by a dislocation moving in the slip plane, we get

$$l_d = \frac{F_{\max}}{\tau b} , \tag{4.15}$$

where $F_{\max} = W/b$ is given by (3.8). Then the number of obstacles to be overcome thermally is given by using q as discussed in Sect. 2.3:

$$f_0 l_d / b = c / q^2 b^2 . \tag{4.16}$$

Equation (4.16) gives the number of effective obstacles, where we have simply neglected the possibility that an effective obstacle has neighbors with characteristic length less than l_d. Inserting (4.14) and (4.15) into (4.16), we have

$$q = (\tau b l_{max}/F_{max})^{1/2} \ . \tag{4.17}$$

In the above consideration, we assumed that no thermal processes occurred before the dislocation had attained the equilibrium configuration consisting of an array of obstacles with characteristic lengths $l_i < l_d$. In other words, the thermal activation processes are assumed to take longer time than the mechanical depinning processes. Furthermore, no lateral motion of the dislocation was allowed on account of the use of the simple distribution function given by (4.16). A more rigorous treatment of these points is given by *Moriya* and *Suzuki* [4.32].

The activation energy corresponding to the thermal process required to surmount the above stronger obstacles using q given by (4.17) can be expressed as

$$U = \frac{2}{3}W - \tau_q b^2 q y \quad \text{for} \quad \tau_q > (\tau_q)_0 \ , \tag{4.18}$$

where the suffix q indicates that the unzipping effect is taken into account, $(\tau_q)_0$ is the plateau stress, and y is given by (3.14). This is the same as the empirical equation (3.22). Putting $\dot\gamma = \dot\gamma_{ext}$, the yield stress can be given by

$$(\tau_q(T))_c^{3/2} = \frac{1}{b^2 y}(F_{max}/b l_{max})^{1/2}\left(\frac{2}{3}W - k_B T \ln A\right) \ , \tag{4.19}$$

where $A = b^2 n \nu_D(x/y)/\dot\gamma_{ext}$. If we take the yield stress for $q = 1$ as $\langle\tau\rangle_T$ and that at $0\,$K as τ_S, (4.19) can be rewritten as

$$(\tau_q(T))_c = 0.68(\mu b^3/W)^{1/6}\tau_S^{1/3}\langle\tau\rangle_T^{2/3}(l_0/l_{max})^{1/3} \ . \tag{4.20}$$

τ_S and $\langle\tau\rangle_T$ are given by (3.21 and 19), and $l_0^2 = b^2/c$. If l_0/l_{max} is an insensitive factor with respect to c, $(\tau_q)_c$ is proportional to $c^{1/2}$ and $T^{2/3}$. Accordingly, it should be noticed that the yield stress is proportional to $T^{2/3}$ even in the case where the rectangular force-distance relation holds (see footnote in Sect. 3.1.3).

Figure 4.12 shows $(\tau_q)_c/\tau_S$ as a function of temperature, where we used values such as $(\mu b^3/W)^{1/6} = 1.34$ for Cu-Ni alloys and $l_0/l_{max} = 10^{-2}$. Since l_0 is $10^{-7}\,$cm for $c = 10^{-2}$, this gives $10^{-5}\,$cm for l_{max}, which seems to be reasonable. The curve is almost linear except for very low temperatures, and it agrees quantitatively with experimental results.

4.4.2 Effects of Inertia on the Activation Process

The enhancement factor Y introduced by *Grananto* for the yield stress in the superconducting state as well as in the temperature region below T_p should be replaced by $Y^* = 1 + \alpha e^{-z}$ for the yield stress at finite temperatures, where $z = \pi\beta/\omega_0$. The value of α is found as follows: Taking the time average of the

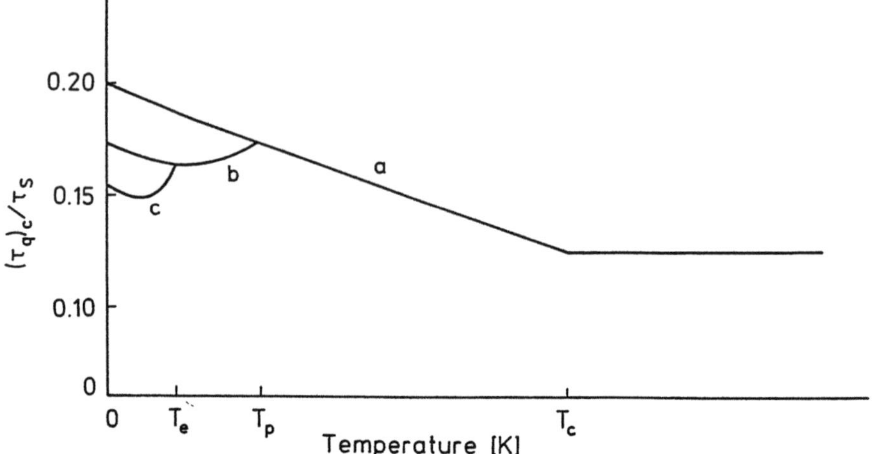

Fig. 4.12. Theoretical curves of yield stress vs temperature. Curve a is obtained from (4.20) by using the values such as $(\mu b^3/W)^{1/6} = 1.34$ and $l_0/l_{\max} = 10^{-2}$ appropriate for Cu-Ni alloy. Curves b and c are schematically drawn taking account of the inertial effects due to the phonon frictional force and the electron frictional force, respectively. Needless to say, curve c appears only in superconducting metals and alloys

displacement ξ in Fig. 4.9 over the period of the characteristic time t^* defined by $e^{-\beta t^*} = 1/e$ gives

$$\alpha = \int_0^{t^*} e^{-\beta t} dt/t^* \ . \tag{4.21}$$

Thus we obtain $\alpha = 0.63$. As a result, the activation energy in terms of the zigzag model is given by, instead of (4.18),

$$U = \frac{2}{3}W - \tau_q b^2 Y_{\mathrm{p}}^* y \qquad \text{(phonon)} \ , \tag{4.22a}$$

$$U = \frac{2}{3}W - \tau_q b^2 Y_{\mathrm{p}}^* Y_{\mathrm{e}}^* y \quad \text{(phonon + electron)} \ , \tag{4.22b}$$

where τ_q is the external stress given by (4.20), the suffix c being omitted, that is, the unzipping effect is taken into account in τ, and accordingly y is the one given by (3.14). In deriving (4.22), we assumed that no intervention of the inertial effects occurs in the unzipping effect.

Schematic curve a in Fig. 4.12 is calculated by using (4.20) taking the unzipping effect into account, curve b represents the inertial effect caused by the decrease of the phonon frictional force (4.22a) and curve c that produced by the decrease of the electron frictional force below the transition temperature from the normal to the superconducting state (4.22b).

The critical coefficient of the phonon friction $(B_{\mathrm{p}})_{\mathrm{c}}$ corresponding to the onset of loss of strength is found from (4.7) and (4.8) by inserting y in place of L to be

$$(B_{\mathrm{p}})_{\mathrm{c}} = 2A\omega_0 \quad \text{where} \quad \omega_0 = \frac{2\pi(C/A)^{1/2}}{y} \ . \tag{4.23}$$

Form this equation we find $(B_{\mathrm{p}})_{\mathrm{c}}$ is proportional to $c^{1/2}$ because y is inversely proportional to $c^{1/2}$ as given by (3.12). In this connection, T_{p} extends only over a very narrow region of temperature with the variation of c so that it may be taken that B_{p} varies approximately in proprotion to T. Accordingly, the peak temperature, T_{p} should be proportional to $c^{1/2}$, as found by experiments shown in Fig. 4.6.

4.4.3 Quantitative Analysis of $\Delta\tau_{\mathrm{ns}}$

The experimental fact that $\Delta\tau_{\mathrm{ns}}$ is proportional to $1 - \Gamma_{\mathrm{BCS}}$ as illustrated in Fig. 4.3 can be theoretically explained as follows. For deformation at a constant strain rate, it holds that $U_{\mathrm{s}}/k_{\mathrm{B}}T = U_{\mathrm{n}}/k_{\mathrm{B}}T$. Put the upper limit of the integral in (4.21) as $1/\beta_{\mathrm{s}}$. Then we get

$$\alpha_{\mathrm{n}} = (B_{\mathrm{s}}/B_{\mathrm{n}})\alpha_{\mathrm{s}} \ . \tag{4.24}$$

Accordingly, $\Delta\tau_{\mathrm{ns}}$ is obtained by using (4.22b) after a short calculation where the suffix q is omitted by simplicity:

$$\Delta\tau_{\mathrm{ns}} = \frac{\xi_{\mathrm{s}}}{1+\xi_{\mathrm{s}}} \left(1 - \frac{B_{\mathrm{s}}}{B_{\mathrm{n}}}\right)\tau_{\mathrm{n}} \ , \tag{4.25}$$

where $\xi_{\mathrm{s}} = \alpha \exp(-z_{\mathrm{s}})$. Equation (4.25) can be rewritten as

$$\Delta\tau_{\mathrm{ns}} = \frac{\lambda\xi_{\mathrm{s}}}{1+\xi_{\mathrm{s}}}(1 - \Gamma_{\mathrm{BCS}})\tau_{\mathrm{n}} \ , \tag{4.26}$$

where $\lambda = (B_{\mathrm{e}})_{\mathrm{n}}/[(B_{\mathrm{e}})_{\mathrm{n}} + B_{\mathrm{r}}]$, and B_{r} is the frictional coefficient due to the bremsstrahlung as discussed in the footnote in Sect. 2.4. This gives $\lambda = 0.6$ for $B_{\mathrm{r}} = (B_{\mathrm{e}})_{\mathrm{n}}/2$. Since it follows that $\xi_{\mathrm{s}}/(1 + \xi_{\mathrm{s}}) \simeq \alpha_{\mathrm{s}}/(1 + \alpha_{\mathrm{s}}) = 0.38$, we have from (4.26)

$$\Delta\tau_{\mathrm{ns}}(0) = 0.23\tau_{\mathrm{n}} \quad \text{at } 0\,\mathrm{K} \ ,$$

which agrees with the value obtained by the experiment described in Sect. 4.1.2.

4.4.4 Effect of Inertia on the Activation Volume

The activation volume is defined by

$$V = \left[\left(k_{\mathrm{B}}T\frac{\Delta \ln \dot{\gamma}}{\Delta\tau_q}\right)(\Delta\tau_q/\Delta\langle\tau\rangle)\right]_T \tag{4.27}$$

or expressed as

$$V = [-(\Delta U/\Delta\tau_q)(\Delta\tau_q/\Delta\langle\tau\rangle]_T \ . \tag{4.28}$$

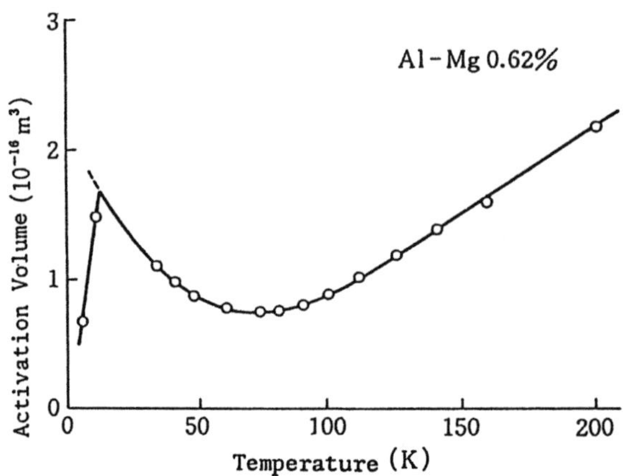

Fig. 4.13. Activation volume as a function of temperature in Al-Mg at.% alloy [4.11]

According to the quasistatic treatment ignoring dynamical effects such as the inertial effect experienced by moving dislocations, the activation volume ($V_0 = b^2 y$) will be independent of temperature, as seen from (3.18). In reality, however, the activation volume increases with decreasing temperature below T_p in normal-conducting alloys, as illustrated in Fig. 4.13. A curve of activation volume versus temperature, obtained experimentally on Al-Mg 0.62 at.% alloy, is plotted in this figure [4.11]. Similar results were obtained for Cu-Al, Cu-Si and Cu-Ag alloys by *Basinski* et al. [4.23].

We obtain the activation volume from (4.20) and (4.22a), using (4.28) as

$$V = A\langle\tau\rangle_T^{-1/3} Y_p^* V_0 , \qquad (4.29)$$

where A is a proportionality constant. From this equation the activation volume is expected to increase with decreasing temperature below T_p on account of the Y_p^* factor as shown in Fig. 4.13. Tendencies to a rapid decrease of V near 0 K and a slow increase above T_p are given by the $\langle\tau\rangle_T^{-1/3}$ factor in (4.29). It is remarked that B_p rapidly drops to almost zero at about 30 K, as shown in Figs. 2.15 and 2.16, giving rise to a peak of V as observed, where Y_p^* must take the value of the upper limit of Y_p equal to 2 as predicted by *Granato*, see (4.9).

5. Dislocation Motion
Controlled by the Peierls Mechanism

Reflecting the periodicity of the crystal lattice, any dislocation in any crystal has its own intrinsic resistance to glide, because the self-energy of the dislocation changes with the periodicity of the lattice. This periodic potential energy with respect to the dislocation position is called the *Peierls potential*, after the researcher who first estimated this potential theoretically for a simple model [5.1]. The stress necessary for the dislocation to surmount this potential, without the aid of thermal energy, is called the Peierls stress. This chapter is concerned with theoretical treatments of the dislocation glide controlled by the Peierls potential.

5.1 Introduction

Since the Peierls potential originates in the periodicity of the crystal lattice, it is naturally influenced strongly by the characteristics of the crystal structure. In the same crystal, the Peierls potential differs from one type of slip system to another. Also, the Peierls potential is considered to reflect strongly the bonding nature of the crystal, because the migration of a dislocation is necessarily accompanied by a bond exchange process at the core. Thus, dislocations in the same slip system in crystals with the same crystal structure and with a similar bonding character will have similar Peierls potentials. The most reliable method to estimate, experimentally, the value of the Peierls stress is to extrapolate the critical resolved yield stress to absolute zero temperature. Table 5.1 lists the experimentally estimated Peierls stresses, normalized with respect to the shear modulus, for various kinds of crystals.

Peierls considered the Peierls potential theoretically by using a simple model where only the two atomic planes abutting on the slip plane were treated as a discrete lattice, the outer parts being approximated by a continuum. This model is called the Peierls model. He assumed a simple cosine function for the interaction potential for the relative displacement of the two lattice planes and obtained the following expression [already given as (2.1)] for the Peierls stress of an edge dislocation [5.2]:

$$\tau_{\mathrm{p}} = \frac{2\mu}{1-\nu} \exp\left(-\frac{2\pi a}{(1-\nu)b}\right) . \tag{5.1}$$

Here, b is the magnitude of the Burgers vector, a the lattice spacing of the glide plane, μ the shear modulus, and ν the Poisson ratio. The Peierls model is an

Table 5.1. Approximate Peierls stress normalized with respect to the shear modulus μ for various kinds of crystals, estimated experimentally

	Slip systems	Examples	τ_p/μ
fcc metals	$\{111\}\langle110\rangle$	Cu, Al, Au, Ag, Pb	$< 10^{-5}$
hcp metals	$(0001)\langle11\bar{2}0\rangle$	Zn, Mg, Cd	$< 10^{-5}$
bcc metals	$\{110\}\langle111\rangle$, $\{112\}\langle111\rangle$	K, Fe, Mo, Ta, Nb	$\sim 5 \times 10^{-3}$
Group IV semiconductors	$\{111\}\langle110\rangle$	Ge, Si	10^0
III-V compounds (zinc blende)	$\{111\}\langle110\rangle$	GaAs, GaP, InAs	10^{-1}
II-VI compounds (zinc blende)	$\{111\}\langle110\rangle$	CdTe, ZnSe	10^{-2}
II-VI compounds (wurtzite)	$(0001)\langle11\bar{2}0\rangle$	CdS, ZnO	10^{-2}
NaCl-type ionic crystals	$\{110\}\langle110\rangle$	LiF, NaCl, KCl	$(1 \sim 5) \times 10^{-4}$
CsCl-type ionic crystals	$\{110\}\langle001\rangle$	CsI, CsBr	$< 10^{-5}$

extremely simplified one and hence the above result cannot strictly be applied to the real crystals. However, the equation signifies qualitatively an important fact, namely, the Peierls stress is very sensitive to the value of a/b. For $a/b = 1$, $1/2$ and $1/3$, $\tau_p = 3.6 \times 10^{-4}\mu$, $3.2 \times 10^{-2}\mu$ and $1.4 \times 10^{-1}\mu$, respectively, assuming $\nu = 0.3$. Consequently, in a crystal the glide system with the largest a/b value, i.e. with the smallest b and largest a, is dominantly active over the other systems. Comparing the dislocations in crystals with different crystal structures, we see from the above equation that in those crystals in which no glide system can take a large a/b value, any dislocation necessarily possesses a high Peierls stress. Therefore, dislocations in crystals with a large unit cell with a complex structure generally possess high Peierls stress, because any slip system in such crystals must have a large Burgers vector and a narrow lattice plane. Metallic crystals generally have large a/b values because of their close-packed structure which is inherent in metallic bonding. Besides metallic crystals, only a limited group of crystals can take an a/b value larger than 0.5; those are NaCl-type crystals, tetrahedrally coordinated crystals such as the diamond structure and the zincblende structure, etc. The main reason why nonmetallic crystals cannot have a high deformability and be utilized as structural materials is that the Peierls stress in these crystals is so high that the dislocations are inactive at stresses below the fracture stress.

In the next section, we discuss the theoretical treatment of the thermally activated glide of a dislocation surmounting the Peierls potential; this type of glide mechanism is called the *Peierls mechanism*. In Chap. 6, we review the plasticity of bcc metals, which are the typical metallic material governed by the Peierls mechanism. Chapter 7 describes the dislocation glide in semiconducting crystals, where some interesting phenomena characteristic of semiconductors occur because the dislocation motion is influenced by the electronic state of the crystal.

5.2 Dislocation Glide by the Peierls Mechanism

A dislocation on a glide plane has a low energy when it lies along a low-index direction in the plane. Namely, the hills and the valleys of the Peierls potential extend in a particular direction, as shown in Fig. 5.1. There are in general several low-index directions in a glide plane; we must have similar wavy potentials in other directions. However, we should note that we cannot express the Peierls potential by a meshed potential as the sum of the wavy potentials in various directions, because the Peierls potential is not a unique function of the position on the glide plane but a function also for the dislocation direction. The Peierls potential is defined only when the direction of the dislocation is given.

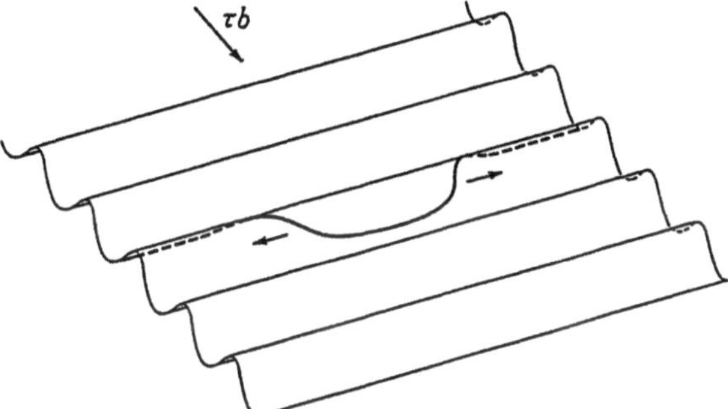

Fig. 5.1. The process of dislocation motion surmounting a Peierls potential by thermal activation

The process a Peierls potential being surmounted by a dislocation lying along a potential valley at a nonzero temperature is as follows. A part of the dislocation surmounts the Peierls potential by thermal fluctuation to form two kinks with opposite signs (kink-pair formation as shown in Fig. 5.1) and then these kinks migrate along the dislocation line (kink migration). In the kink migration process, the kink energy also changes periodically, reflecting the periodicity of the lattice in the direction of the dislocation line. The potential barrier for the kink migration is called the *Peierls potential of the second kind.*

The processes of kink-pair formation and kink migration are theoretically classified into two cases. In one case, the height of the Peierls potential E_p is much smaller than the self-energy of the dislocation line E_s which includes the Peierls potential. In this case, the line tension approximation can be applied to the kink because the width of a kink is much wider than the atomic spacing and the shape of the kink is smooth enough (see the following section). As a result, any part of the kink is nearly parallel to the Peierls potential valley and hence the local energy at any point of the kink can be approximated by the potential for the straight dislocation. Furthermore, since the Peierls potential of the second kind for a smooth kink is negligibly small compared with that of the first kind, the thermally activated motion of the dislocation is controlled only by the kink-pair formation process. The other case is that E_p occupies a considerable part of E_s, the width of the kink is on the order of an atomic distance and the Peierls potential of the second kind is on the same order as that of the first kind. For a dislocation severely curved in an atomic dimension, the line tension model is no longer applicable and thus the dislocation motion must be treated based on the kink model. The former case is called the smooth kink model and the latter the abrupt kink model.

5.2.1 Smooth Kink Model

The problem of thermally activated kink-pair formation was first treated theoretically by *Seeger* [5.3] as a mecahnism for a low-temperature internal friction peak called the Bordoni peak, and since then the theory has been modified and improved. Figure 5.2 shows schematically the process of kink-pair formation. The dashed line 1 in the figure indicates the initial stable position of the dislocation and the dashed line 2, the next stable position. Suppose that a part of the dislocation is bowed out by thermal fluctuation. When the bow-out is only slight as in (a) in the figure, the segment will return to the original stable position, but when it is to the extent shown in (d), the two kinks formed will be separated far apart by the forces exerted by an applied stress, and eventually the whole dislocation will be translated to position 2. Thus, there must be a saddle point in between these two configurations. The energy necessary to bow out the dislocation to the saddle point is the activation energy for kink-pair formation. As the applied stress is increased, the effective Peierls potential, which includes the work-done

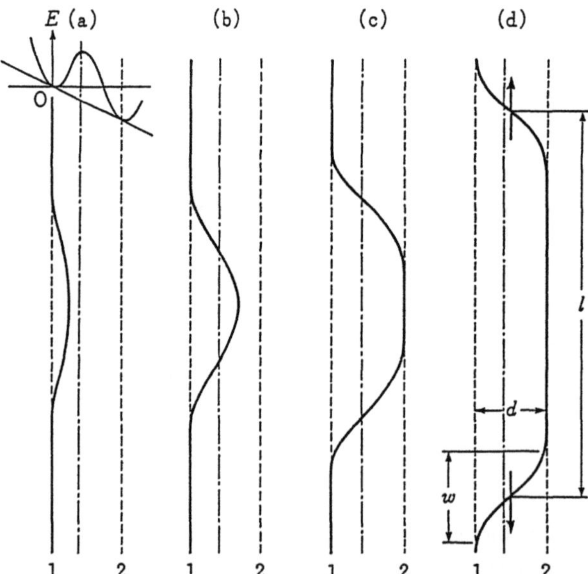

Fig. 5.2a–d. Schematics of kink-pair formation in the smooth kink model. Broken lines 1 and 2 show the Peierls potential valleys and dot-and-dash lines, the potential hills

by the applied stress, becomes lower and the maximum of the potential closer to the initial stable position. As a result, the saddle point configuration becomes like (b) or (a). At low stress, on the other hand, the two kinks must be separated widely, as in (d), before the saddle point is reached. In such a case, the kink-pair formation energy can be obtained from the kink-kink interaction energy [5.4].

a) Kink-Pair Formation Energy in a Low Stress Range. The energy of a single kink[1] E_k is given by [5.5]

$$E_k \sim \frac{2d}{\pi}\sqrt{2E_p E_0} = \frac{2\sqrt{2k}d}{\pi}E_0 \;, \tag{5.2}$$

where E_0 is the line energy of the dislocation per unit lenght, E_p the height of the Peierls potential, k the ratio E_p/E_0, and d the period of the Peierls potential. When the separation of the two kinks l is much larger than the kink width w ($\sim d\sqrt{E_0/2E_p} = d/\sqrt{2k}$), the interaction energy between the two kinks is given by [5.5]

$$E_{int} = -\frac{K}{8\pi}\frac{\mu b^2 d^2}{l} \;, \tag{5.3}$$

where K is a constant equal to $(1-2\nu)/(1-\nu)$ for the edge dislocation and $(1+\nu)/(1-\nu)$ for the screw dislocation. Thus, the energy (i.e. the increment of energy with respect to the straight dislocation) of a kink-pair with a separation l is expressed, taking into account the work-done term $\tau b l d$, as

$$E = \frac{4\sqrt{2k}\,d}{\pi}E_0 - \frac{K}{8\pi}\frac{\mu b^2 d^2}{l} - \tau b l d \;. \tag{5.4}$$

This energy has a maximum value

$$E_{kp} = \frac{4\sqrt{2k}\,d}{\pi}E_0 - \sqrt{\frac{K}{2\pi}}\,bd\sqrt{\mu b d\tau} \tag{5.5}$$

for

$$l = l^* = \frac{1}{2}\sqrt{\frac{K}{2\pi}}\sqrt{\frac{\mu b d}{\tau}} \;. \tag{5.6}$$

E_{kp} given by (5.5) is the activation energy of kink-pair formation in a low stress range.

b) Kink-Pair Formation Energy at High Stress. At high enough stress, the saddle point configuration is only a bulge of a part of the dislocation as in Fig. 5.2b, and thus the treatment by two kinks is no longer valid. Kink-pair formation in such a case was discussed independently by *Dorn* and *Rajnak* [5.6], and *Celli* et al. [5.7].

---

[1] E_k is equal to a half of the kink-pair formation energy at zero stress, see Sect. 5.2.1b

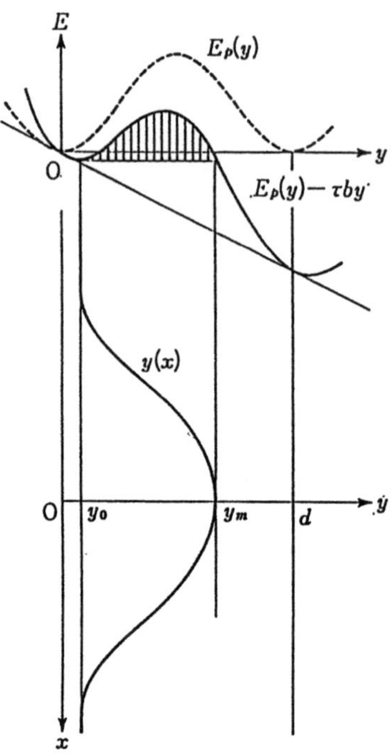

Fig. 5.3. Saddle point configuration of kink-pair formation at high stress

We can obtain the saddle point energy by taking account of the fact that the dislocation configuration at the saddle point is in an unstable equilibrium state. In the upper part of Fig. 5.3, we show the unstressed Peierls potential by a dashed line and that under stress by a solid line. The initial equilibrium position of the dislocation before the bow-out is at y_0, which satisfies $-dE/dy = \tau b$. The total energy increment for a bow-out configuration is expressed by the following equation as a sum of the dislocation line energy, the Peierls potential and the work done:

$$E(y(x)) = \int_{-\infty}^{\infty} \left\{ [E_0 + E_p(y)]\sqrt{1 + y'^2} \right.$$

$$\left. -[E_0 + E_p(y_0)] - \tau b(y - y_0)\right\} dx .$$ (5.7)

From the conditions $y' \equiv dy/dx \ll 1$ and $E_p(y) \ll E_0$, Eq. 5.7 is approximated as

$$E(y(x)) \simeq \int_{-\infty}^{\infty} \left[\frac{1}{2}E_0 y'^2 + E_p(y) - E_p(y_0) - \tau b(y - y_0)\right] dx .$$ (5.8)

Since at the saddle point every part of the dislocation is in an equilibrium state, the equation

$$E_0 y'' - \frac{dE_p(y)}{dy} + \tau b = 0$$ (5.9)

must be satisfied due to the line tension approximation. After integrating (5.9) with respect to y, we obtain

$$y'^2 = \frac{2}{E_0} \left\{ [E_p(y) - \tau by] - [E_P(y_0) - \tau by_0] \right\} , \tag{5.10}$$

using the relations $y'' = (1/2)dy'^2/dy$ and $y' = 0$ at $y = y_0$. At the top position of the bow-out, i.e. at y_m in the figure, $y' = 0$, and hence from (5.10) y_m satisfies

$$E_p(y_m) - \tau by_m = E_p(y_0) - \tau by_0 . \tag{5.11}$$

It is thus found that y_m corresponds to a position at which the potential value $E_p(y) - \tau by$, shown by a solid line in Fig. 5.3, takes the same value as the initial minimum position. After substituting (5.10) in (5.8) and changing the integral variable from x to y, we obtain the saddle point energy as

$$\begin{aligned}
E_{kp} &= 2E_0 \int_{y_0}^{y_m} y' dy \\
&= 2\sqrt{2E_0} \int_{y_0}^{y_m} \sqrt{[E_p(y) - \tau by] - [E_p(y_0) - \tau by_0]} \, dy .
\end{aligned} \tag{5.12}$$

The value in the root in the integrand corresponds to the vertical bar drawn in the $E_p(y) - \tau by$ curve in Fig. 5.3. By use of the above equation, the relations between E_{kp} and τ have been calculated for various functional forms of $E_p(y)$.

The kink-pair formation energy at zero stress, E_{kp}^0, is the sum of the two energies of isolated kinks. For a Peierls potential with a cosine function with an amplitude of $kE_0/2$, i.e.

$$E_p(y) = \frac{kE_0}{2} \left(1 - \cos\frac{2\pi y}{d} \right) , \tag{5.13}$$

E_{kp}^0 can be calculated as

$$E_{kp}^0 = \frac{4\sqrt{2k}\, d}{\pi} E_0 \tag{5.14}$$

from (5.12). This is actually twice the value of E_k in (5.2). For the cosine Peierls potential of (5.13), $\tau_p = (\pi k/db)E_0$. Using $E_0 \sim \mu b^2$ and $d \sim b$, $\tau_p \cong \pi k\mu$. Thus, putting $\tau_p = \beta\mu$ we find from (5.14) that E_{kp}^0 and β can be related by

$$E_{kp}^0 = \frac{4\sqrt{2\beta}\, d}{\pi^{3/2}} E_0 \cong \sqrt{\beta b^3}\, \mu . \tag{5.15}$$

It should be noted here that even though E_{kp} (5.12) gives the correct value at $\tau = 0$, it contains an error for $\tau \gtrsim 0$, because (5.12) does not contain the kink-kink interaction energy, which should be taken into account.

c) **Asymptotic Form Near τ_p.** Since τ_p corresponds to the slope at the inflection point of the $E_p(y)$ curve, the functional form near τ_p can be approximated by a

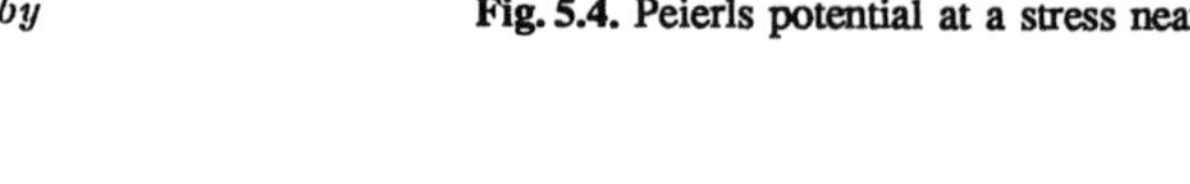

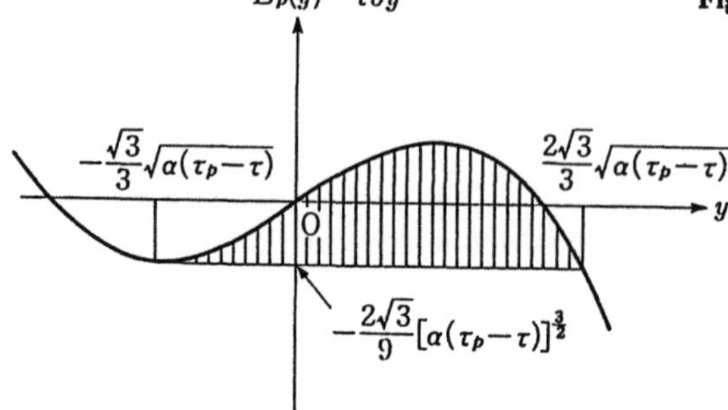

third-order polynomial of y. If we express the functional form of $E_p(y) - \tau b y$ as $-y^3 + \alpha(\tau_p - \tau)y$, the potential shape is like that shown in Fig. 5.4. From (5.12)

$$E_{kp} \propto \int_{-p}^{2p} \sqrt{-y^3 + \alpha(\tau_p - \tau)y + 2p^3}\, dy$$

$$= \int_{-p}^{2p} \sqrt{(y + p)^2 (2p - y)}\, dy \ ,$$

where $p = (\sqrt{3}/3)\sqrt{\alpha(\tau_p - \tau)}$. After evaluating the integral, we obtain [5.9]

$$E_{kp} \propto (\tau_p - \tau)^{5/4} \ . \tag{5.16}$$

This result is in contrast to the result $E \propto (\tau_c - \tau)^{3/2}$ for the point obstacle case (see footnote in Sects. 3.1.2 and 4.4.1), and is a distinctive feature of the Peierls mechanism.

5.2.2 Dislocation Velocity in the Smooth Kink Model

As mentioned before, since the Peierls potential (of the second kind) for the kink motion is small, once the kink-pair formation takes place, kinks move at a high velocity. Consequently, the dislocation velocity is controlled only by the kink-pair formation. The existence of impurities, may retard the kink motion and hence reduce the dislocation velocity [5.7]. When the impurities interact strongly with the dislocation, they may even affect the kink-pair formation energy [5.10–12]. In this section, however, we treat the dislocation motion ignoring impurity effects.

From the results of the previous subsections, E_{kp} for the whole range of the stress is obtained as shown in Fig. 5.5. In the low stress region the $E_{kp} - \tau$ curve tends to a parabola l_1 expressed by (5.5). Approximating $d \sim b$, $E_{kp} = 2E_k - K'b^3\sqrt{\mu\tau}$, where the constant K' is about 3 for a screw dislocation and 7 for an edge dislocation. At higher stress, the curve is expressed by a line l_2 obeying (5.12), and near τ_p, $E_{kp} = K''(\tau_p - \tau)^{5/4}$ according to (5.16). In the transition region from the low stress regime to the high stress one, the $E_{kp} - \tau$

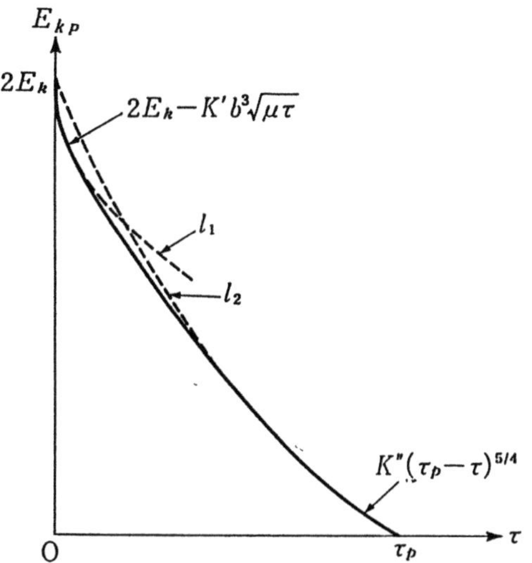

Fig. 5.5. Stress dependence of the activation energy of kink-pair formation of the whole stress range

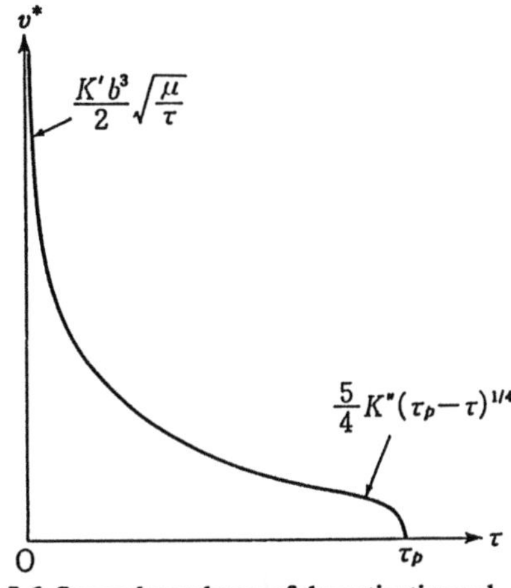

Fig. 5.6. Stress dependence of the activation volume of kink-pair formation

curve is expected to interpolate smoothly from the curve l_1 to l_2. From this $E_{kp}-\tau$ relation, we can calculate the activation volume defined by $v^* = -(\partial E_{kp}/\partial\tau)$; the v^* versus τ relation is schematically drawn in Fig. 5.6. For $\tau \to 0$, $v^* \sim (K'b^3/2)\sqrt{\mu/\tau}$ and for $\tau \to \tau_p$, $v^* \sim (5/4)K''(\tau_p - \tau)^{1/4}$.

For a kink-pair formation energy E_{kp}, the frequency at which the kink-pair formation takes place per unit length of the dislocation is expressed from rate theory as

$$\nu_{kp} = \frac{1}{l}\nu_l \exp\left(-\frac{E_{kp}}{k_B T}\right) . \tag{5.17}$$

Here l is the kink-pair separation at the saddle point and ν_l the vibrational frequency of the dislocation segment with length l; the reciprocal, $1/l$, corresponds to the number of sites for kink-pair formation. ν_l is approximately equal to $\nu_D(l/b)$, where ν_D is the Debye frequency. Let L be the distance of kink motion. Then the dislocation velocity is given by

$$v = \frac{dbL\nu_D}{l^2} \exp\left(-\frac{E_{kp}}{k_B T}\right) . \tag{5.18}$$

L is determined by the length of the dislocation or the separation between strong obstacles such as junction points of the dislocation with forest dislocations. It should be noted here that although in a low stress range the frequency of the kink-pair formation on the opposite side of force becomes appreciable, under the condition that the kink migration is athermal, no kink migration against the force can occur. Consequently, the temperature dependence does not become $\sinh(E/k_B T)$, which results from the contribution of the reverse process.

We will now estimate the order of magnitude of the pre-exponential factor $v_0 \equiv dbL\nu_D/l^2$. We take the value of l as twice the kink width to an approximation, though l is actually a decreasing function of the stress. For $\tau_p \sim 0.01\mu$ as in bcc metals, the kink width $w \sim 10b$. Let $d \simeq b$, $L \simeq 0.1$ mm and $\nu_D \simeq 10^{13}\,\text{s}^{-1}$, then v_0 is calculated to be $\sim 2 \times 10^6$ m/s. The shear strain rate $\dot{\gamma}$ is given by

$$\dot{\gamma} = \varrho bv = \varrho bv_0 \exp\left(-\frac{E_{kp}}{k_B T}\right) , \tag{5.19}$$

where ϱ is the mobile dislocation density. For $\varrho = 10^{11}\,\text{m}^{-2}$ and $\dot{\gamma} = 10^{-4}\,\text{s}^{-1}$, the value of the exponent in (5.19) is calculated to be -27. This value agrees well with the values obtained experimentally for bcc metals (described later).

5.2.3 Abrupt Kink Model

In the case where the Peierls stress is as high as $\mu/10$, the kink width calculated based on the line tension approximation is $\sim 4d$ and the kink energy according to (5.2) is $\sim 0.2E_0 d$. Usually, the energy of the aprupt kink is nearly equal to the core energy of the dislocation segment of lenght d, i.e. of the order of $0.1E_0 d$ [5.5]. In these circumstances, for high Peierls potential it is highly probable that the kink part lies in the valley of the Peierls potential of the second kind, forming an abrupt kink. In particular, when the Peierls potential is determined by the energy required to break atomic bonds at the core, as in semiconducting crystals, the Peierls potential of the second kind is comparable with that of the first kind, and thus the kink shape must be abrupt. In this subsection, we treat the thermally activated motion of a dislocation with abrupt kinks.

The processes of the dislocation motion are: an elementary kink-pair is first formed on a straight dislocation (Point A in Fig. 5.7) and then these kinks migrate along the dislocation line, successively overcoming the Peierls potential of the second kind E_p^k. The thermal-equilibrium concentration of kinks in a unit length of the dislocation under zero stress is given by

$$c_k = c_k^+ + c_k^- = \frac{2}{d'} \exp\left(-\frac{E_k}{k_B T}\right) \tag{5.20}$$

where c_k^+ and c_k^- are respectively the concentrations of plus and minus kinks

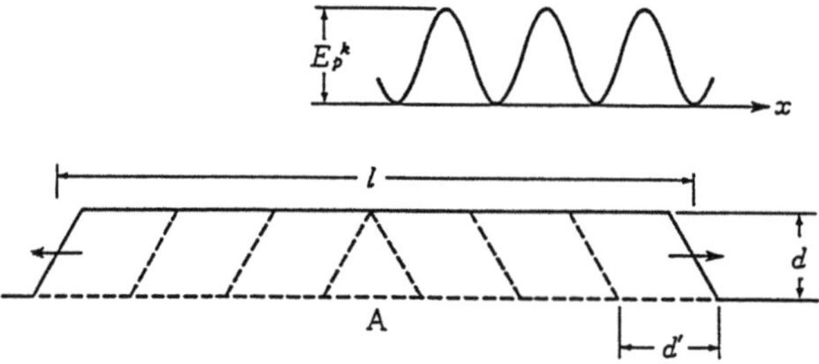

Fig. 5.7. Process of kink-pair formation in the abrupt kink model

and d' is the period of Peierls potential of the second kind. Thermal equilibrium is realized by the balance between the rate of formation of thermally activated kink pairs and the mutual annihilation of plus and minus kinks by the diffusional motion of them along the dislocation line. The thermal equilibrium state is perturbed under the action of the stress because the plus kinks and minus kinks drift in opposite directions. At high enough temperatures and low stress, however, the kink diffusion is faster than the kink drift velocity, and hence plastic deformation proceeds, keeping the kink concentration approximately equal to the equilibrium concentration (5.20). In this case, the average position of a kink, randomly walking back and forth, drifts in the direction of the force, which brings about plastic deformation. Meanwhile, at low temperatures and high stress, kinks migrate dominantly in the direction of the force, and the steady-state kink concentration can no longer be expressed by (5.20); the dislocation velocity is governed by the steady-state condition determined by the rate of kink-pair formation and the kink migration velocity.

a) High Temperature and Low Stress Case. The diffusion coefficient of a kink is given by

$$D_k = \nu_k d'^2 \exp\left(-\frac{E_p^k}{k_B T}\right) , \tag{5.21}$$

where ν_k is the vibrational frequency of a kink, approximately equal to the Debye frequency ν_D. At low stress, using the Einstein relation [drift velocity] = [mobility: $D/k_B T$] $\times$ [force], the drift velocity of a kink, v_k, is given by

$$v_k = \frac{D_k}{k_B T}\tau bd = \frac{\tau bdd'}{k_B T}\nu_k d' \exp\left(-\frac{E_p^k}{k_B T}\right) . \tag{5.22}$$

The condition that must be satisfied in order that the kink concentration can be approximated by the thermal equilibrium concentration c_k is that the time of pair annihilation of kinks due to drift motion ($\sim 1/c_k v_k$) is much longer than the lifetime of kinks determined by the thermal diffusivity ($\sim 1/c_k^2 D_k$). From (5.20–22) we obtain

$$\frac{\tau bdd'}{2k_B T} \ll \exp\left(-\frac{E_k}{k_B T}\right) , \tag{5.23}$$

for the above condition.[2] For example, for $E_k \approx 1\,\mathrm{eV}$, $k_B T = 0.1\,\mathrm{eV}$ ($T = 1200\,\mathrm{K}$) and $d = d' = b = 3\,\mathrm{\mathring{A}}$, $\tau \ll 5\,\mathrm{MN\,m^2}$ from (5.23). When the above condition is satisfied, the dislocation velocity is given by the following equation from (5.20) and (5.22):

$$v = dc_k v_k = \frac{2\tau bd^2 d'}{k_B T}\nu_k \exp\left(-\frac{E_k + E_p^k}{k_B T}\right) . \tag{5.24}$$

---

[2] A more sophisticated computation by *Kawata* and *Ishioka* [5.13] showed that (5.24) is approximately valid for a rather wide stress range.

b) Relatively High Stress Case. At a relatively high stress, the velocity of kink motion is determined by the difference in the frequency between the forward jump and the backward jump, i.e.

$$v_k = d'\nu_k \left[\exp\left(-\frac{E_p^k - \tau b d d'/2}{k_B T} \right) - \exp\left(-\frac{E_p^k + \tau b d d'/2}{k_B T} \right) \right]$$

$$= 2d'\nu_k \exp\left(-\frac{E_p^k}{k_B T} \right) \sinh\frac{\tau b d d'}{2k_B T} \; . \tag{5.25}$$

When $\tau b d d'/2k_B T \ll 1$ is satisfied, the above equation naturally coincides with (5.22). To obtain the steady-state dislocation motion, we must know the frequency of the kink-pair formation. Figure 5.8 shows schematically the energy change in the process of kink-pair formation. The abscissa indicates the separation between kinks. The monotonically increasing broken line in the figure is equal, except near the origin, to $2E_k$ minus the kink-kink interaction energy given by (5.3). Adding the Peierls potential for the kink migration to this, we obtain a wavy potential curve. Under a stress τ, adding the work-done term $-\tau b l d$, we obtain the potential curve as drawn by a solid line in the figure. The lower envelope curve of the potential gives the following maximum value equivalent to (5.5) at $l = l^*$ given by (5.6):

$$E_{kp} = 2E_k - \sqrt{\frac{K}{2\pi}} \, b d \sqrt{\mu b d \tau} \; . \tag{5.26}$$

The frequency of the kink-pair formation in such a potential is determined by the rate at which the generated kink pair diffuses out beyond the critical position l^*. The theory of thermally activated kink-pair formation accompanied by the diffusional motion of kinks has been treated by *Lothe* and *Hirth* [5.5, 14] and *Seeger* and *Schiller* [5.15].

For Peierls stress of the order $\mu/10$ and μ, E_p^k is estimated to be $\sim 0.02\,E_0$ and $\sim 0.2\,E_0$, respectively, and hence the ratio $E_p^k/2E_k$ is 0.1 and 1, respectively,

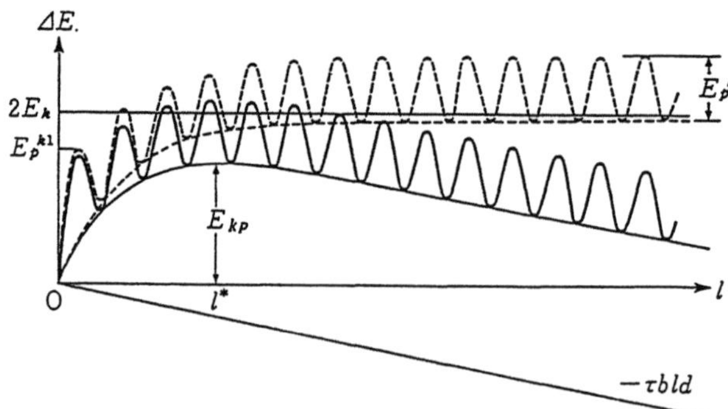

Fig. 5.8. Energy change in the process of kink-pair formation in the abrupt kink model. l is the separation between kinks

since $2E_k \simeq 0.2E_0$. Thus, we find that for crystals in which the Peierls stress is of the order of μ, like Si, $2E_k$ and E_p^k are of similar magnitude.

The frequency of the kink-pair formation for the potential shown in Fig. 5.8 can be obtained in the following way. At $\tau = 0$, the number density of kinks with $\Delta E(l)$ in unit length of the dislocation is [5.5]

$$\varrho_k(\Delta E) = c_k^+ c_k^-(l) = \frac{1}{d'^2} \exp\left(-\frac{\Delta E(l)}{k_B T}\right) . \tag{5.27}$$

Since the work-done term in the region $l^* < l$ is much smaller than E_{kp}, (5.27) holds approximately even under a stress. Thus, the density of kink pairs with separations around l^*, c_{kp}, is expressed by

$$c_{kp} = \varrho_k(E_{kp})d' = \frac{1}{d'} \exp\left(-\frac{E_{kp}}{k_B T}\right) . \tag{5.28}$$

The frequency of kink-pair formation ν_{kp} corresponds to the flux of the kink-pair flow from l^* to a larger value of l. Using the result of (5.25), we obtain

$$\nu_{kp} = c_{kp}\frac{v_k}{d'} = \frac{2v_k}{d'} \exp\left(\frac{E_p^k + E_{kp}}{k_B T}\right) \sinh\frac{\tau b d d'}{2 k_B T} . \tag{5.29}$$

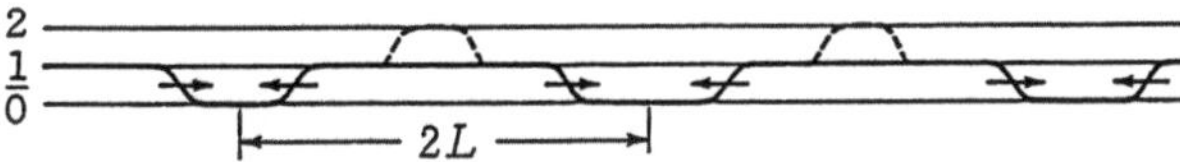

Fig. 5.9. A schematic drawing showing the steady motion of a dislocation, where the rate of kink-pair formation and that of annihilation of plus and minus kinks are balanced

The steady state of the dislocation motion is determined by the condition that during a lifetime (from the generation of a kink pair to the annihilation with opposite kinks after migrating a mean free path L) one new kink pair is formed on the average in a region of the dislocation with length $2L$, see Fig. 5.9. The mean lifetime of a kink pair is L/v_k. Considering that during a lifetime the average dislocation length in the new potential valley (position 1 in Fig. 5.9) is L, the steady-state condition is written as

$$\nu_{kp} L \frac{L}{v_k} = 1 . \tag{5.30}$$

It follows then that

$$L = \sqrt{\frac{v_k}{\nu_{kp}}} = \sqrt{\frac{d'}{c_{kp}}} = d' \exp\left(\frac{E_{kp}}{2 k_B T}\right) . \tag{5.31}$$

In covalent crystals, E_{kp} is typically $2\,\text{eV}$, and hence the value of L is $\sim 5\,\mu\text{m}$ at $T = 1200\,\text{K}$ and $10\,\text{cm}$ at $T = 600\,\text{K}$ from (5.31). In the former case, the steady state of dislocation motion may be realized with a steady-state kink density, whereas, in the latter case, since the kink mean free path is much larger than the dislocation length Λ (or the separation between strong obstacles such as intersecting forest dislocations) the above steady state cannot be realized. For $2L \ll \Lambda$ the dislocation velocity is expressed from (5.29) and (5.31) by

$$v = 2dL\nu_{kp} = 4d\nu_k \exp\left(-\frac{E_{kp}/2 + E_p^k}{k_B T}\right) \sinh\frac{\tau bdd'}{2k_B T} . \tag{5.32}$$

For $2L \gg \Lambda$, we obtain

$$v = d\Lambda\nu_{kp} = \frac{2d\Lambda}{d'}\nu_k \exp\left(-\frac{E_{kp} + E_p^k}{k_B T}\right) \sinh\frac{\tau bdd'}{2k_B T} . \tag{5.33}$$

In real materials at a particular temperature, it is not possible to distinguish, from the velocity measurements alone, which of the two equations (5.32) and (5.33), controls the velocity, because they have the same functional form. To obtain, experimentally, the mean free path of the kink motion, it is necessary to observe the dependence of the dislocation segment length on the dislocation velocity. Because for $2L \gg \Lambda$, v is proportional to Λ and for $2L \ll \Lambda$, it is independent of Λ, the critical segment length above which the dislocation velocity becomes almost constant should correspond to $2L$. A successful attempt has been made to measure the segment length dependence of the dislocation velocity for a silicon single crystal in which forest dislocations had been introduced to control the segment length [5.17].

At $\tau = 10^{-3}\mu$, ($\sim 100\,\mathrm{MN\,m^2}$), the value of $\tau bdd'$ in (5.32) and (5.33) is of the order of 0.01 eV, and hence even at room temperature $\tau bdd'/2k_B T \sim 1/5$. As a result, in actual deformation conditions, $\sinh(\tau bdd'/2k_B T)$ can be approximated by $\tau bdd'/2k_B T$. Thus, both (5.32) and (5.33) are expressed approximately by the same functional form,

$$v = v_0 \frac{\tau b^3}{k_B T} \exp\left(-\frac{E}{k_B T}\right) . \tag{5.34}$$

It should be noted here that the energy necessary to produce the first unit kink pair at a point A in Fig. 5.7 is obviously different from the potential barrier for the subsequent kink migration. In cases where the Peierls potential of the first kind is much higher than that of the second kind, E_p^{k1} in Fig. 5.8 can be as high as $2E_k$. In such a case, the rate of kink-pair formation is determined dominantly by E_p^{k1}, and hence we should use the following expression as the value of ν_{kp} (5.30–33) instead of (5.29):

$$\nu_{kp} = \frac{\nu_D}{d'} \exp\left(-\frac{E_p^{k1}}{k_B T}\right) . \tag{5.35}$$

So far, we have treated the problem with the condition that the stress range is much lower than τ_p. The τ_p value in crystals where the dislocation motion must be treated by the abrupt kink model is generally so high that the crystals cannot be deformed without fracture at a stress near τ_p. Consequently, the observed dislocation motion is well described with the present approximation.

6. Dislocations in bcc Metals and Their Motion

Copper (and its alloys) and iron or steel form two groups of metals that have been utilized since the ancient times. More recently, aluminium and its alloys have found many applications, thus forming three important metallic groups. When we compare the low-temperature flow stresses of these three kinds of metals, we find that that of Fe is much higher than the other two: at liquid nitrogen temperature usual steels can no longer be plastically deformed without fracture. From the fact that Fe has body-centered cubic (bcc) structure while Cu and Al have face-centered cubic (fcc) structure, such a difference in low-temperature plasticity is considered to be due to the difference in crystal structure. In fact, while all pure fcc metals exhibit weak temperature dependence of the yield stress, metals with bcc structure show, without exception, a strong temperature dependence. In this chapter, we describe characteristic plastic behavior of bcc metals and its interpretation.

6.1 Dislocations in bcc Metals and Their Peierls Potential

Concerning the origin of the strong temperature dependence of the yield stress of bcc metals, there had long been two opposing interpretations. One, the extrinsic theory, is that the cause is solid solution hardening due to interstitial impurities, C, N, etc., while the intrinsic theory holds that the origin is an inherent high Peierls potential of dislocations in bcc metals. Advances in experimental research have shown that the latter is the correct explanation for the following reasons: (i) In experiments on highly purified bcc metals, the strong temperature dependence of the flow stress does not change even if the impurity concentration is extrapolated to zero. (ii) In in situ transmission electron microscopy of plastic deformation in bcc metals at low temperature, steady, continuous motion of straight screw dislocations is observed as expected from the Peierls mechanism. (iii) Computer simulation studies show that the Peierls stress for the screw dislocation is much higher than that of the edge dislocation in bcc crystals (consistent with experiments, as will be described in the subsequent subsection). (iv) Characteristic orientation dependence of slip behavior in single crystals at low temperature is difficult to understand in terms of impurity effects while it is interpretable by the Peierls mechanism based on the computer simulation studies.

Although it is now established that the low temperature plasticity of bcc metals is controlled by an intrinsic resistance to the screw dislocation motion, since the 1960s there have still been two different views of the origin of the intrinsic resistance. One view is based on the threefold symmetric dissociation of the screw dislocation. The possibility of the existence of the stacking fault in the {112} plane with the fault vector (1/6)⟨111⟩ had long been considered in relation to the coherent twin boundary in bcc metals. If the stacking fault is actually stable, there is a possibility that the screw dislocation with the Burgers vector (1/2)⟨111⟩ has the lowest energy when it is dissociated into three partial dislocations with the Burgers vector (1/6)⟨111⟩, forming stacking faults on threefold-symmetric {112} planes, as illustrated in Fig. 6.1. In this dislocation configuration, the dislocation motion must be preceded by the constriction of partials in order for the dislocation to glide without dragging stacking faults, just like the cross slip process of dissociated dislocations in fcc metals. This means that the dislocation glide always requires constriction energy, resulting in immobilization of the dislocation at low temperatures. Such a dislocation dissociation is named the sessile dissociation and the dislocation glide mechanism accompanying the constriction is often called the pseudo-Peierls mechanism.

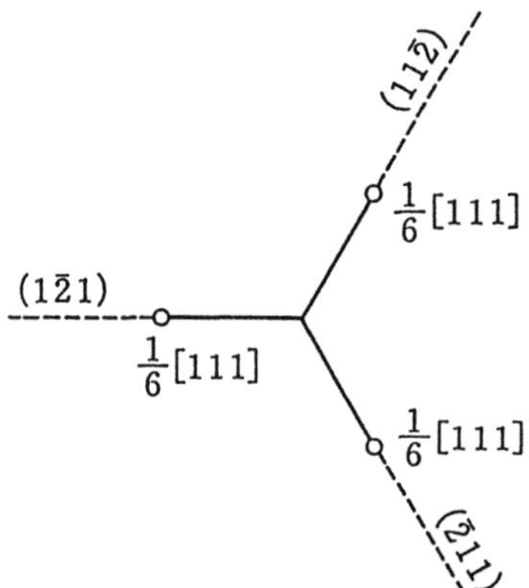

Fig. 6.1. Sessile dissociation model of a screw dislocation in the bcc lattice with $b = (1/2)\langle 111 \rangle$

Actually, however, there is no experimental evidence that the dislocation in bcc metals is dissociated as given in Fig. 6.1, and also computer simulation studies do not show the existence of the stable stacking fault. Meanwhile, computer simulations of the core structure of the screw dislocation have often indicated that the strain field is not isotropic but threefold symmetric with a pattern similar to that in Fig. 6.1. Due to such results, there still remains the concept of the pseudo-Peierls mechanism where the threefold symmetric spreading of the core strain is the cause of the immobile nature of the screw dislocation in bcc crystals, although the distinct dissociation model as presented in Fig. 6.1 has now been discounted.

Suzuki [6.1] was the first to point out that a screw dislocation with $b = (1/2)\langle 111 \rangle$ in any bcc lattice must have high energy positions and low energy

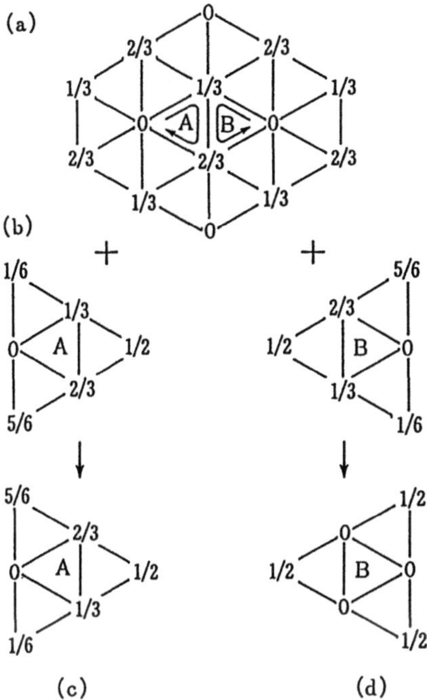

Fig. 6.2a–d. Changes of the atomic positions when a screw dislocation is introduced at position A or B in the perfect lattice of (**a**). Numerals indicate atomic positions (in the direction perpendicular to the paper) of atomic rows in units of the Burgers vector

positions due to the geometrical features of the lattice itself, which is the origin of the high Peierls potential. Figure 6.2a shows the bcc lattice viewed from a $\langle 111 \rangle$ direction where the numerals indicate atomic positions, in the direction perpendicular to the paper, in units of the Burgers vector. It is seen from this figure that the atoms in neighboring triangular $\langle 111 \rangle$ atomic rows form a spiral configuration and its winding direction changes alternately as shown by A and B in Fig. 6.2a. If we superimpose the strain field of the screw dislocation shown in (b) on the perfect lattice, the atomic positions are altered as shown in (c) or (d) for a dislocation positioned at A or B, respectively. Atomic positions of the three central atomic rows in (c) take the same spiral configuration as in the perfect lattice, whereas those in (d) come to the same level, resulting in a reduction of the interatomic distance by about 6%. Consequently, the core energy of the dislocation at B is considered to be higher than that of A. Computer calculations using interatomic potentials support this conclusion based on simple geometrical considerations. Thus, the screw dislocation in the bcc lattice has minimum energy positions at triangular positions with a separation of $(2\sqrt{2}/3)b$ and Peierls potential barriers between these positions. The motion of the dislocation in this case can be regarded as controlled by the usual Peierls mechanism.

Which of the two concepts, the Peierls mechanism or the pseudo-Peierls mechanism, is more appropriate to describe the motion of the screw dislocation in bcc crystals? In the following section, we describe the computer simulation studies of the core structure of the screw dislocation and its motion under stress, and discuss the results in relation to the above problem.

6.2 Computer Experiments

The estimate of the Peierls stress based on the Peierls approximation mentioned in Sect. 5.1 is useful for qualitative understanding, e.g. of the dependence of the value of a/b on the Peierls stress, but too crude to apply to individual crystals. On the other hand, to analyze the core structure and the Peierls stress for real crystals from first principles is too difficult in practice, because the number of atoms relevant to the problem is too great. Thus, we have to rely on computer simulation studies of model crystals. Since the end of the 1960s, with the development of high-speed computers, such studies have become popular (especially for bcc crystals) [6.2].

The most important problem in atomistic simulation studies, not only of dislocations but also of other lattice defects, is how to choose the interaction potentials between atoms. Although it has now become possible to derive the cohesive energy of simple crystals with considerable accuracy from first principles based on electron theory, it is still impossible to apply such a theory to large lattice defects such as a dislocation. Therefore, we have to rely on an approximate interatomic potential, often called the semi-empirical potential. Usually, central force potentials are used which stabilize the crystal structure and reproduce the elastic properties of the crystal.

Of the various kinds of bcc metals, alkali metals can best be approximated by the central force potential; the lattice energy can be expressed approximately by the central force potential determined by the pseudopotential theory, ignoring the volume-dependent energy term because of the small volume change accompanying introduction of the dislocation. For transition bcc metals for which a number of experiments have been performed, semi-empirical potentials have been applied, but the energy of the lattice in these crystals cannot be well expressed by the central force potential alone because the cohesive energy of these metals involves an orientational bonding due to d-electrons. Consequently, it is more appropriate to interpret the obtained results as those of fictituous bcc lattices rather than real bcc metals.

6.2.1 Crystal Geometry and Peierls Stress

We can naturally take into account properly the effect of the crystal geometry on the Peierls stress in the computer simulation, but not in the Peierls model. *Basinski* et al. [6.3] computed the Peierls stresses for screw dislocations in bcc and hcp Na using the pseudopotential for Na. They showed that the Peierls stress for the bcc lattice is more than 25 times higher than that for the hcp lattice for the same potential. *Vitek* and *Yamaguchi* computed the Peierls stresses for the screw dislocation [6.4] and the edge dislocation [6.5] using the so-called Johnson-type potential, proposed by *Johnson* [6.6], and obtained 10–20 times higher Peierls stress for the screw dislocation than the edge dislocation. These facts suggest that the high Peierls stress of the screw dislocation in bcc lattices originates in the crystal geometry, as mentioned in the preceding section.

Considering that the core structure and the Peierls stress are rather sensitively dependent on the interatomic potential assumed, and that the potential does not guarantee us to reproduce a real crystal, it is necessary to change the potential systematically and to see the general tendency of the effect of the potential on the results. The first trial along this line was performed by *Suzuki* for bcc lattices by use of interatomic row potentials [6.7]. The results of computer simulation conducted for similar models [6.8–10] will be summarized below.

6.2.2 Core Structure of a Screw Dislocation

According to the linear elasticity theory, the strain field of the screw dislocation has a displacement component only in the direction of the Burgers vector. Also, in the present lattice model, we assume that the energy of the screw dislocation is determined by the relative displacements in the direction of the Burgers vector between atomic rows parallel to it. When we view the bcc lattice from the direction of the Burgers vector $(2/1)\langle 111\rangle$, the $\langle 111\rangle$ atomic rows form a triangular arrangement as shown in Fig. 6.2. The toal lattice energy is now assumed to be described by the sum of interaction energies between adjacent atomic rows.

We express the interatomic row potential between a pair of $\langle 111\rangle$ atomic rows, shown in the inset to Fig. 6.3, as a function of the difference of the atomic positions, z, parallel to the rows, as schematically shown in the figure. $\phi(z)$ is a periodic function with period b and is symmetric with respect to $z = b/2$. The curvature of the potential at $z = b/3$, which corresponds to the perfect lattice position, is related to the shear modulus of the perfect crystal by $\phi''(b/3) = \mu/\sqrt{3}$. The maximum of the potential is usually at $z = nb$, because the interatomic distance becomes shortest at these positions. The minimum of the potential is always at $z = b/2$ when we convert any interatomic potentials proposed so far for bcc metals into interatomic row potentials. This comes from the fact that interatomic potentials which stabilize the bcc lattice have a minimum at a position between the first and the second nearest neighbors. Thus, we postulate that the interatomic row potential has a shape like that shown in Fig. 6.3.

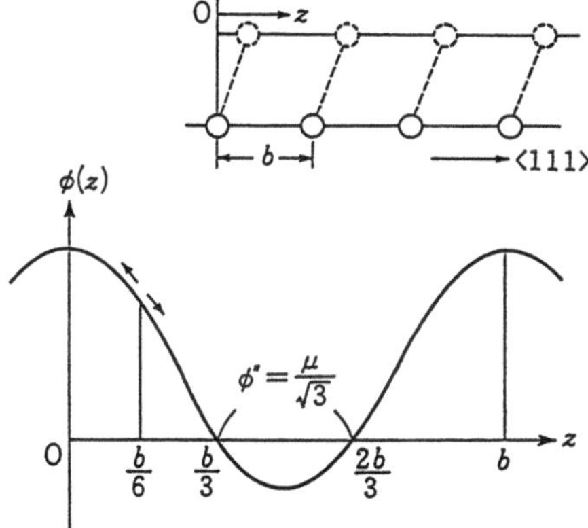

Fig. 6.3. Interaction potential between $\langle 111\rangle$ atomic rows in the bcc lattice

In our interatomic row potential model expressing the energy of a screw dislocation, there is inevitably a drawback in that the effect of the volume change, or the displacements of atoms perpendicular to the atomic rows (which occur at the very center of the screw dislocation core), cannot be taken into consideration. Nevertheless, our model possesses important advantages. Firstly, systematic simulation studies for systematically varying potentials are possible, because the computation time is much shorter for our model than for interatomic potential models. Secondly, and more importantly, we should note that the structure of the dislocation core and the Peierls stress are determined essentially by the shape of the interatomic row potential rather than interatomic potential; if the two interatomic row potentials which are converted from two different interatomic potentials are similar, then the dislocation core structure and its behavior under stress are similar.

Two series of different functional forms of $\phi(z)$ in Fig. 6.3 have been used; (i) cosine Fourier series up to second order, and (ii) polynomial functions, both with varying parameters. As a result, three types of stable core structures under zero stress have been obtained, as shown in Fig. 6.4. In this way of representing the core structure, devised by *Vitek* et al. [6.11], arrows between atomic rows indicate relative displacements of atomic positions, with respect to the perfect lattice, in the direction of the atomic row. The magnitude of an arrow fully connecting two neighboring rows corresponds to a displacement of $b/3$, and hence the total sum of the arrows of any closed circuit amounts to three times the interatomic row distance when the circuit includes a dislocation center, and zero when it does not.

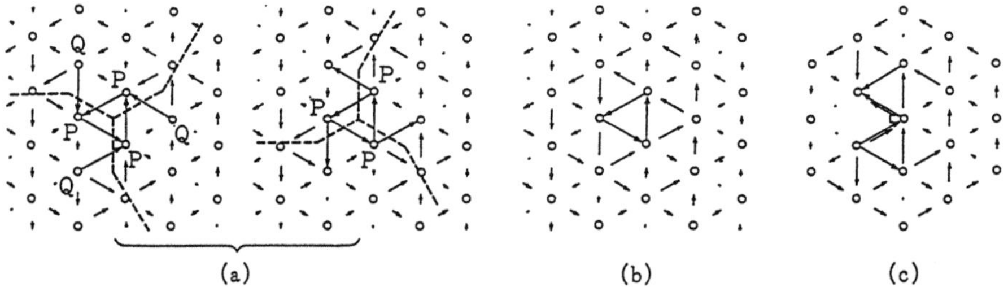

(a)   (b)   (c)

Fig. 6.4a–c. Three types of dislocation core in the bcc lattice obtained by computer simulation [6.10]: (a) polarized type, (b) isotropic type and (c) split type

The center of the dislocation in (a) and (b) is located at the type-A site in Fig. 6.2, while that in (c), determined by the strain field far from the core, is near an atomic row (the center is indicated by a square in the figure) and the core spreads over two type-A sites. The strain field of the type (b) core is isotropic and close to that obtained by elasticity theory. Compared with the type (b) core, the atomic configurations of the type (a) core are such that the three atomic rows indicated by P's displace upward or downward (the left and the right configurations in different directions) by about $b/10$, and the next outer three rows (Q's) displace in the opposite directions by about half the displacements of the P's. The two configurations of Fig. 6.4a coincide with each other under $180°$ rotation with its axis perpendicular to the dislocation line, and hence are energetically equivalent, so the type (a) core is often called the degenerate type.

The degeneracy of the core structure was fist pointed out by *Vitek* [6.12]. *Seeger* and *Wüthrich* called this type of core the polarized core, from the characteristic displacements of the central atomic rows mentioned above [6.13]. In the following description of the core behavior under stress, we prefer to call the type (a) core the polarized core, because under a stress the degeneracy breaks. The polarized core has a characteristic threefold-symmetric spreading of the strain field as shown by dashed lines in the figure. We will call the type (b) core the isotropic-type core and the type (c) core the split-type core.

For a particular interatomic row potential, either the polarized-type or the isotropic-type core becomes-stable. The split-type core appears usually as a metastable configuration when the stable configuration is of the isotropic type. Even in the polarized core, the split core sometimes becomes stable under the action of a stress, as described below.

What characteristic features of the potential determine whether the stable core is the isotropic core or the polarized core? The atomic positions at the center of a dislocation with an isotropic core are those given in Fig. 6.2c, i.e. the three central atomic rows are located at $z = \pm b/6$ with respect to the next three atomic rows. A simultaneous displacement of the three central rows changes the interaction energy by $3[\phi(b/6) + \delta) + \phi(b/6 - \delta) - 2\phi(b/6)]$. Thus, as far as the six atom rows are concerned, the sign of $\phi''(b/6)$ determines the stability of the isotropic core. More exactly, we must of course take into account the interactions with outer atomic rows.

A more detailed calculation derives the following conditions of the potential for the isotropic core to be stable [6.14]:

$$A > 0 , \quad C > 0 ,$$
$$D \equiv B^2 - AC < 0 , \tag{6.1}$$

where $A = \phi''(b/6) + \phi''(0.386b)$, $B = \phi''(b/6)$, $C = \phi''(b/6) + \phi''(0.447b) + \phi''(0.294b)$. If the above conditions are not satisfied, the three central rows are displaced and the core type becomes polarized. Approximating $\phi''(0.386b) \simeq \phi''(b/3)$ and $\phi''(0.447b) + \phi''(0.294b) \simeq 2\phi''(b/3)$, the above conditions can be rewritten as

$$\phi'' \left(\frac{b}{6} \right) > -\frac{2}{3} \phi'' \left(\frac{b}{3} \right) = -\frac{2\mu}{3\sqrt{3}} . \tag{6.2}$$

Thus we find again that the curvature at $z = b/6$ plays an essential role. Figure 6.5 shows schematically the two potential curves which yield different core types, (a) the isotropic core and (b) the polarized core.

It is difficult to understand why the split-type core (Fig. 6.4c) becomes metastable for the isotropic core but unstable for the polarized core. Phenomenologically, it seems that the energy difference between the isotropic configuration

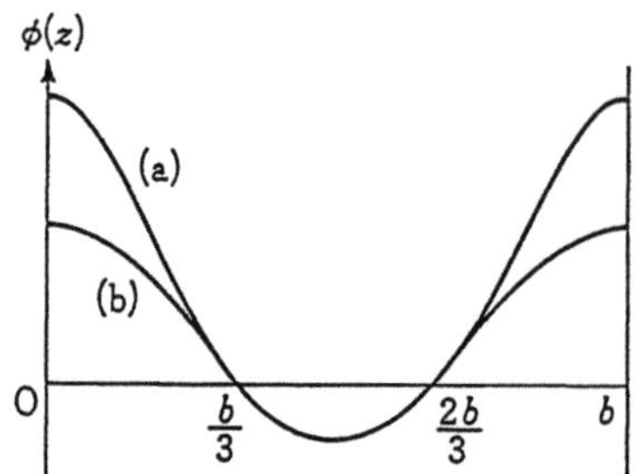

Fig. 6.5. Characteristics in the shape of the interatomic row potential which produces an isotropic core (*a*) or a polarized core (*b*)

and the split configuration is almost constant and hence the polarized core, whose energy is decreased with respect to the isotropic configuration, becomes relatively more stable with respect to the split configuration, which makes the latter configuration unstable.

Most of the computer simulation studies performed up until the early 1970s, which used assumed interatomic potentials, showed that the stable core structure is of the polarized type. This yielded a tendency towards the belief that the threefold-symmetric core spreading in the polarized core plays an essential role in the dislocation behavior. Later, however, it was shown in various models [6.15, 16] that the isotropic core can also exist, and thus an interesting question is which type of core structure real screw dislocations possess.

6.2.3 Behavior Under Stress

Computer experiments on glide of dislocations in model bcc crystals have been conducted since the beginning of the 1970s [6.5, 6, 9, 10, 15–19]. Effects of nonglide stress components, i.e. stress components other than the shear stress component on the slip system, on the glide behavior have been reported [6.5, 19], and they have been shown to be related to elastic anisotropy [6.19]. In the following, we concentrate our attention on the effects of the shear stress component.

In a $\langle 111 \rangle$ zone of a cubic crystal, there are three $\{110\}$ planes and three $\{112\}$ planes, mutually intersecting at every 30° and any particular $\langle 111 \rangle$ zone plane exists in every 30° sector; hence the shear stress plane is represented by a plane in one $\{110\}$-$\{112\}$ sector. However, if we consider the possible effect of the sense of the shear stress in addition to the shear stress plane, we must consider a 60° range, ±30° on either side of a $\{110\}$ plane. This is because, except for the $\{110\}$ plane, the crystal structure is not equivalent for the sense of the shear direction. A typical case is the $\{112\}$ shear shown in Fig. 6.6, where for one shear the crystal can be twinned and for the other it cannot; the former type of shear is called the $\{112\}$ twinning shear (denoted $\{112\}_\text{T}$) and the latter type the anti-twinning shear ($\{112\}_\text{AT}$).

The effects of the sense of the shear stress, in addition to the shear stress plane, on the slip behavior of crystals were first demonstrated experimentally for an Fe-Si alloy [6.20], and later recognized to be a characteristic, common feature of the plasticity of bcc crystals. Customarily, the orientation of a single

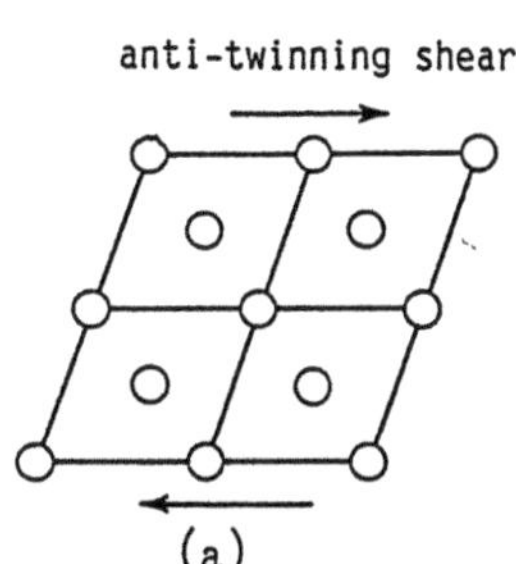

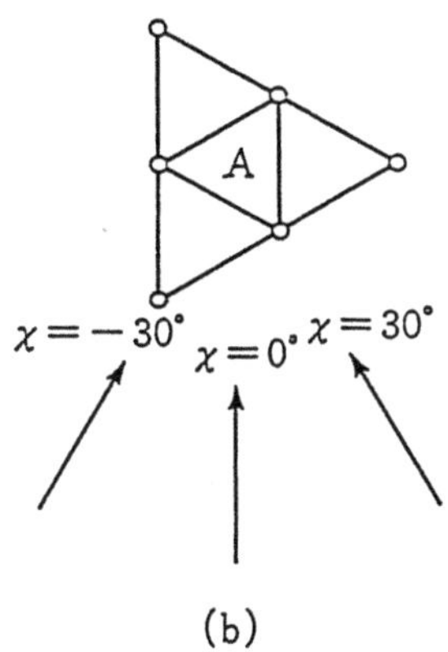

Fig. 6.6. (a) Nonequivalence of the opposite senses of the $\{112\}\langle111\rangle$ shear in the bcc lattice. (b) Representation of the direction of force on a screw dislocation. Position A corresponds to the stable position of a dislocation viewed from the $\langle111\rangle$ direction, see Fig. 6.2

crystal is represented by an angle χ between the maximum shear stress plane (in the $\langle111\rangle$ zone) of the crystal and the nearest $\{110\}$ plane; the sign of the χ value is defined so that $\chi = 30°$ corresponds to the anti-twinning $\{112\}$ shear and $\chi = -30°$ to the twinning $\{112\}$ shear.

In Fig. 6.6b, in which circles indicate $\langle111\rangle$ atomic rows, the directions of the force for $\chi = 0°$, $\chi = \pm30°$ are shown by arrows for the dislocation located at A. For a simple shear stress τ in the slip direction, the resolved shear stress on the $\{110\}\langle111\rangle$ system is $\tau \cos\chi$ and those on the $\{112\}\langle111\rangle$ systems are $\tau \cos(\pm30° - \chi)$. The glide plane of the dislocation is usually expressed, similarly to the shear stress plane, by an angle ψ from the nearest $\{110\}$ plane.

How does the dislocation core behave under stress? The results obtained so far vary greatly from model to model, and cannot simply be classified by the core type alone [6.9, 10, 18]. Nevertheless, we can see typical behavior, characteristic of each core type, isotropic or polarized, which is described below. To illustrate the dislocation motion schematically, we use three kinds of marks representing the three core types, as given in Fig. 6.7. The two types of windmill marks indicate polarized cores with opposite polarizations.

A special feature of the polarized core is that the dislocation glides along the $\{112\}_T$ or $\{112\}_{AT}$ plane, except for a special case where the core strain spreads for a long distance along three $\{110\}$ planes. The typical behavior of the motion of the dislocation core under a stress above the Peierls stress is given in Fig. 6.7a, b. A unit step of a dislocation is a translation from one stable site to any of the three sites, of the six neighboring sites, to which the strain of the core spreads, i.e., in the direction of the blade of the windmill mark in Fig. 6.7. An important consequence of the unit translation is that the polarization is always reversed. As a result, the movement of the core is always a zigzag along a $\{112\}$

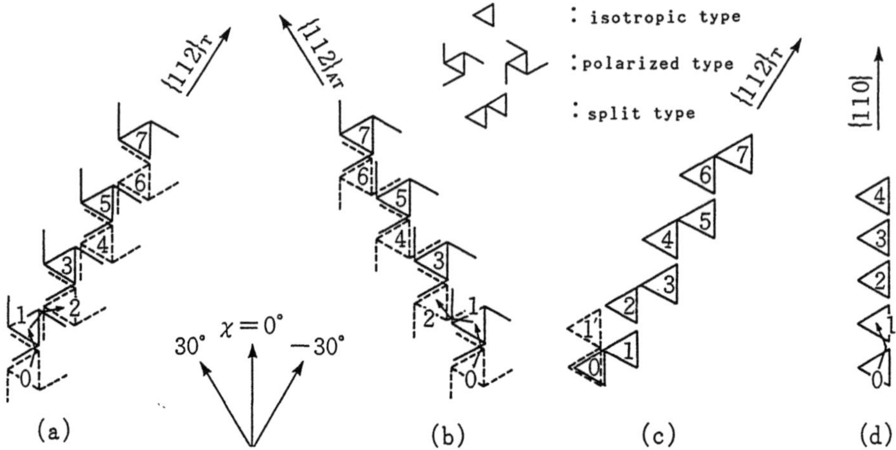

Fig. 6.7a–d. Schematic representation of the movement of dislocation cores under stress in computer simulation studies at absolute zero temperature. (a) and (b) are for a polarized dislocation core, and (c) and (d) for an isotropic core

plane as $0 \rightarrow 1 \rightarrow 2 \rightarrow 3 \ldots$ in the figure. The selection of the glide plane, $\{112\}_T$ or $\{112\}_{AT}$, depends on the stress condition of the χ value. The range of the $\{112\}_T$ glide is much wider than that of the $\{112\}_{AT}$ one. This is because the trace of the dislocation center during the translation from one stable site to another, which is determined from the outer strain field, is not straight but zigzag, as shown by bent arrows in Fig. 6.7. This means that a unit translation path passes through a position of the split core. In Fig. 6.8a is shown an example of the orientation dependence of the Peierls stress[1] and that of the glide plane for this type of core; the lower figure is the $\tau_p - \chi$ relation and the upper one the $\psi - \chi$ relation.

The movement of the dislocation core is thus not simple. A general tendency is that the $\{112\}_T$ glide is most active and τ_p at $\chi = +30°$ is always higher than that at $\chi = -30°$. The above results are for absolute zero temperature. To simulate the motion at nonzero temperature, we must first obtain the Peierls potentials for various translations, calculate transition probabilities for these translations according to the theory of kink-pair formation (mentioned in Sect. 5.2.1) and then compute the average motion of the dislocation from calculated relative translation probabilities [6.21]. As a result of such calculations, we find that at nonzero temperature the dislocation undergoes glide not along a low index plane but along an irrational plane resulting in the so-called noncrystallographic glide. Such a glide appears generally in the case where the slip is controlled by the glide of nondissociated screw dislocations, as in bcc metals.

---

[1] One may consider that the Peierls stress is defined as a unique value, independent of orientation, but this is impossible for the screw dislocation in bcc metals, in which the Schmid law does not hold (Sect. 6.3). The Peierls stress, here, is defined as a function of the orientation of the shear stress plane.

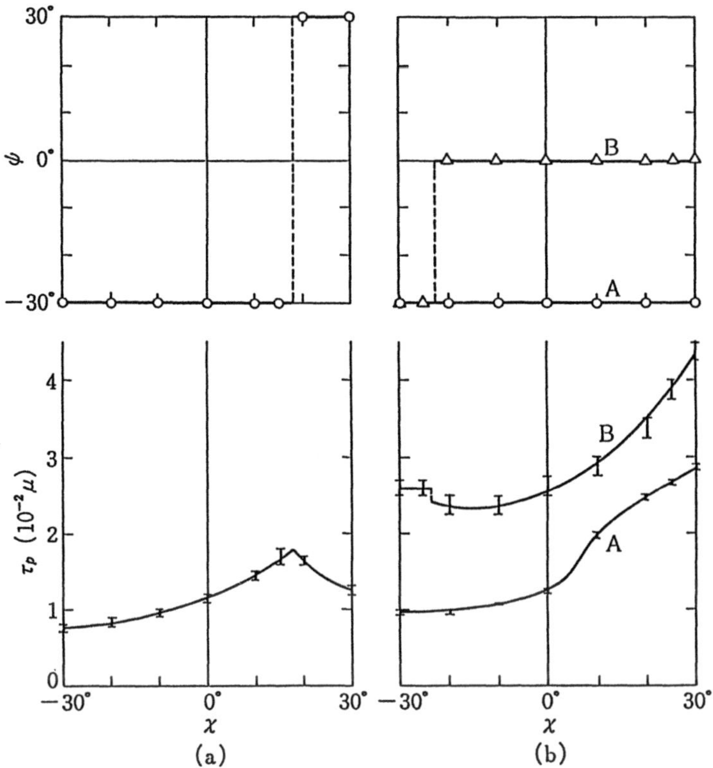

Fig. 6.8a,b. Examples of ψ vs χ and τ_p vs χ at absolute zero temperature obtained in computer simulation studies [6.10]. (a) For a polarized core and (b) for an isotropic core

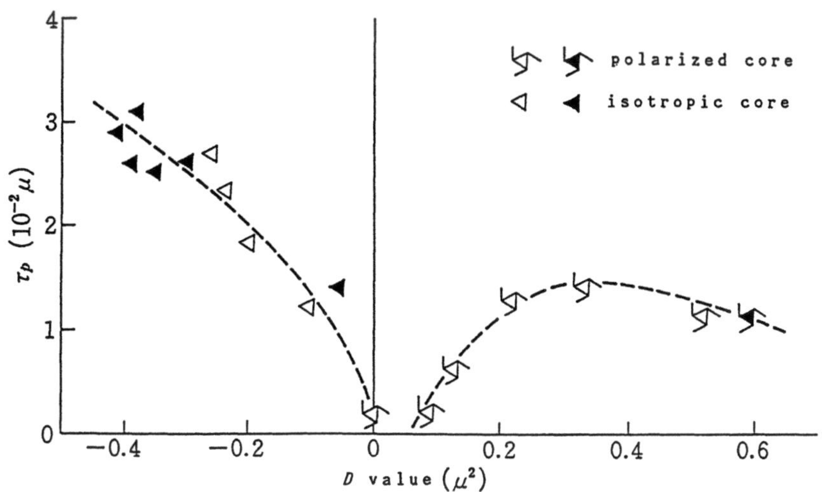

Fig. 6.9. Computed Peierls stresses at $\chi = 0$ for model bcc crystals as a function of the D value (6.1) [6.10]

In cases where the core strain does not distribute in a definite glide plane as in nondissociated screw dislocation, the Peierls approximation is not a good one for estimating the Peierls stress. In the Peierls approximation, the Peierls stress is higher in crystals with higher ideal shear strength [6.22], but for screw dislocations in model bcc crystals no such correlation has been obtained [6.10]. Instead, when we plot the Peierls stresses (at $\chi = 0°$) of model crystals against the D values in (6.1), we obtain a good correlation as shown in Fig. 6.9. The D value indicates the relative stability of the core type between the isotropic core and the polarized core, and hence for a small absolute value of D the stability of the core is weak. This leads to a low Peierls stress. Figure 6.9 also shows that the notion that the high Peierls stresses in bcc crystals originate in the polarized nature of the core is not correct.

6.3 Plasticity of bcc Metals

6.3.1 Yielding of bcc Metals

As mentioned already, recent experiments on high-purity transition metals have revealed their intrinsic plastic properties. The residual resistivity ratio of the high purity specimens used for the plasticity experiments is of the order of 10^4, and the concentration of interstitial impurities estimated from the resistivity ratio is of the order of 1 at. ppm. It has thus become widely believed that the low temperature plasticity of bcc metals is governed basically by the Peierls mechanism, but there still remains some controversy over the role of impurities in the plasticity. One cause of this lies in the ambiguous definition of the yield stress in discussing the low temperature plasticity.

Figure 6.10 shows stress-strain curves for three different specimens. Body-centered cubic metals of low purity generally exhibit an upper yield point fol-

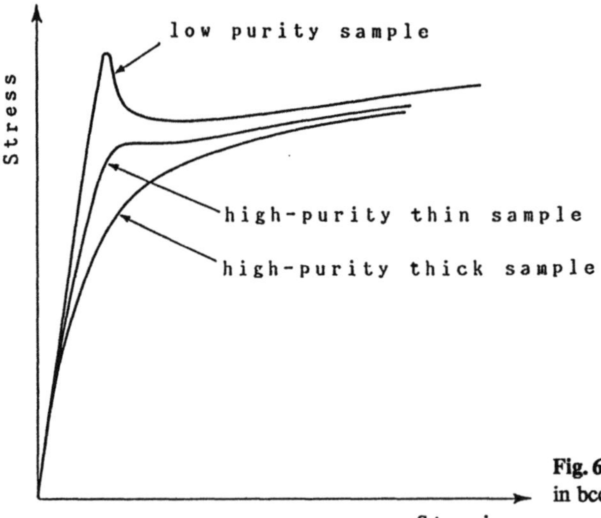

Fig. 6.10. Initial stage of stress-strain curves in bcc metal single crystals

lowed by a yield drop. This phenomenon occurs because grown-in dislocations are locked by impurity segregation and hence multiplication of dislocations can take place only at high stress. High-purity bcc metals, on the other hand, undergo gradual yielding, forming the so-called stage-0. The existence of the stage-0 in high-purity specimens makes the definition of the yield stress equivocal. When the specimen is thin, however, the stress-strain curve bends rather sharply even for a high-purity sample as drawn in the figure. This can be interpreted as follows. As mentioned before, edge dislocations in bcc metals, unless they are locked by impurities, are much more mobile than screw dislocations. As a result, in high-purity specimens yielding takes place first by motion of edge dislocations at a low stress, but soon the specimens are exhaustion-hardened because multiplication cannot occur without the screw dislocation motion. Thus the motion of grown-in edge dislocations produces the stage-0. A quasi-steady deformation involving a multiplication process is realized only at high stress where the motion of screw dislocations becomes active. Let the density of grown-in edge dislocations be ϱ_e and the specimen diameter d, the stage-0 strain γ_e is given by $\sim \varrho_e b d/2$, i.e. γ_e is proportional to the specimen diameter. For $\varrho_e = 10^6\,\mathrm{cm}^{-2}$, γ_e amounts to 1.5% for $d = 1\,\mathrm{cm}$.

Thin specimens serve not only for sharp yielding but also for low temperature experiments in the following two respects. Firstly, the thin specimen suppresses occurrence of twinning and thus prevents brittle fracture [6.23, 24]. Secondly, thin specimens have a much lower tendency to thermal instability deformation, which is one characteristic feature of low temperature plasticity. The thermal instability deformation, which was first demonstrated for Al alloys by *Basinski* [6.25] occurs as a result of specimen heating by plastic work done; a local heating by plastic deformation accelerates the deformation and yields a serrated stress–strain curve. When the ratio of the surface to volume is large, as in thin specimens, the thermal instability deformation tends to be suppressed because the heat released from the specimen surface is relatively large compared with the heat production by plastic deformation. In the following subsection, we describe systematic investigations, at low temperatures down to liquid helium temperature, for high-purity single-crystal specimens of bcc metals. These experiments have successfully been done mainly by a group at Kyushu University, Japan.

6.3.2 Plasticity of bcc Metal Single Crystals

Yield stresses discussed below are those due to screw dislocation motion after the plastic deformation reaches a quasi-steady state. The crystal orientation will be represented by the angle χ, the position of the maximum shear stress plane of the primary slip system, as in the previous section. For a stress axis A in the standard stereographic triangle shown in Fig. 6.11, the sign of χ is positive toward the plane $(\bar{2}11)$ from $(\bar{1}01)$ in tension tests and toward the plane $(\bar{1}\bar{1}2)$ in compression tests.[2] We should note that when the stress axis is close to [001] the

---

[2] In the literature, the sign of χ is often taken to be positive toward $(\bar{2}11)$ irrespective of the testing mode. Here, we follow the definition of the previous section, paying attention only to the sense of the shear stress.

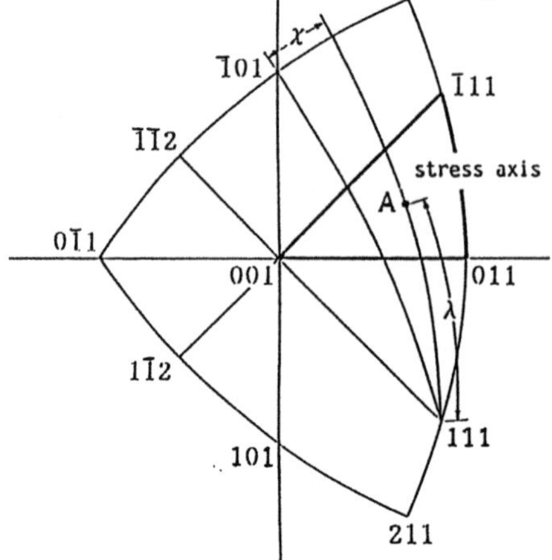

Fig. 6.11. Representation of the orientation of the stress axis

slip direction becomes $[\bar{1}11]$ and thus the angle χ in this case represents the angle from the (101) plane. The observed slip plane orientation is also represented by the angle ψ defined in the same way. The stress τ stands for the shear stress resolved in the maximum shear stress plane, i.e., $\tau = \tau_a \sin\lambda \cos\lambda$, where τ_a is the applied tension or compression stress and λ is given in Fig. 6.11.

a) τ-χ and ψ-χ Relations. Figure 6.12 shows τ_y-χ and ψ-χ relations for various bcc metals at liquid helium temperature (only the data of τ_y for LiMg are at 20 K). τ_y is normalized with respect to the shear modulus. From these results we can classify glide behavior into two cases. In one case the glide plane is always $\{110\}$ and τ_y is minimum near $\chi = 0°$, as in Fe and Mo. In the other

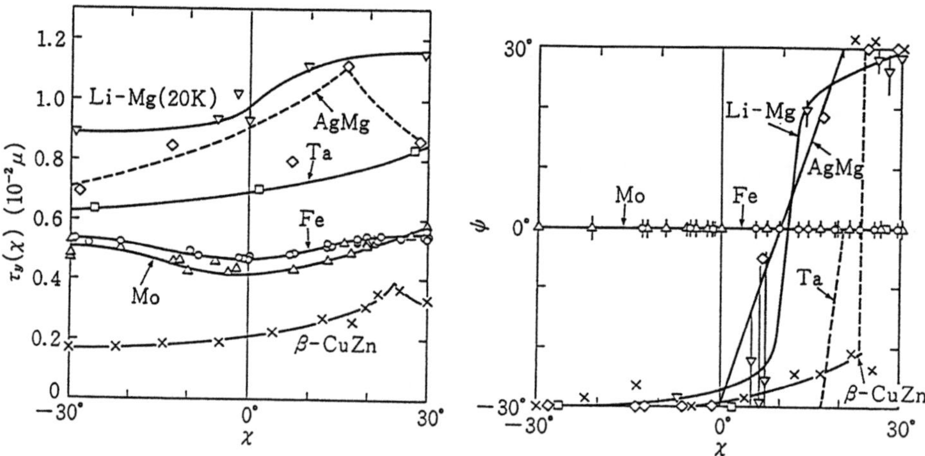

Fig. 6.12. τ_y-χ and ψ-χ relations at helium temperature for various bcc metal single crystals (Fe [6.26], Mo [6.27], Ta [6.28], AgMg [6.29], β-CuZn [6.30], Li-65 at.% Mg [6.31])

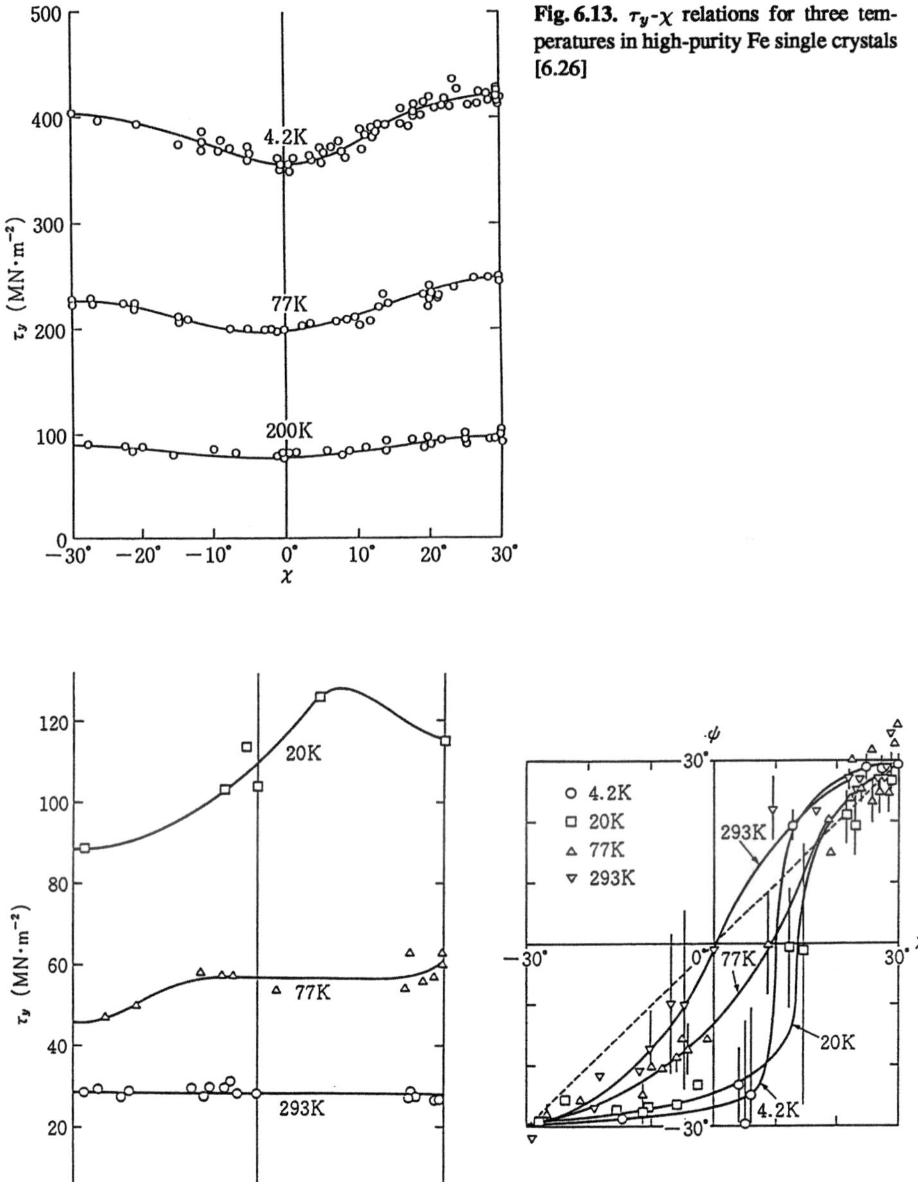

Fig. 6.13. τ_y-χ relations for three temperatures in high-purity Fe single crystals [6.26]

Fig. 6.14. τ_y-χ and ψ-χ relations for Li-65 at.% Mg single crystals at various temperatures [6.31]

case, the glide plane is $\{112\}_T$, except for the orientation near $\chi = 30°$, in which it is $\{112\}_{AT}$, and τ_y is minimum at $\chi = -30°$ and maximum near the orientation at which the glide plane changes. We will name these two cases the $\{110\}$ glide type and $\{112\}$ glide type. It should be noted that these two cases are similar to the behavior seen in the computer simulation for the polarized core (Fig. 6.8(a)) and for the isotropic core (Fig. 6.8(b)). These similarities seem to indicate that the core structures of Fe and Mo are of the isotropic type, and that the others have polarized cores. Incidentally, simulation of the glide behavior of the screw dislocation in CsCl-type ordered lattices reproduces the behavior of β-CuZn and AgMg [6.32].

Figure 6.13 shows experimental τ_y-χ curves for Fe at various temperatures. Figure 6.14 shows τ_y-χ and ψ-χ relations for a Li-Mg alloy. It is seen from these τ_y-χ curves that the orientation dependence of the yield stress becomes weak as the temperature is increased. The ψ-χ relation in Li-Mg shows that the glide plane transition becomes more and more obscure, and at room temperature the glide plane is near the maximum shear stress plane in any orientation. For high-purity iron, no ψ-χ curve has been obtained owing to the diffuse nature of glide bands at higher temperatures, but the results for Fe alloys exhibit similar tendencies to the Li-Mg alloy, i.e., the glide plane deviates from the $\{110\}$ plane more and more toward the maximum shear stress plane as the temperature is increased.

Such a noncrystallographic glide at higher temperature originates from the fact that the contribution of thermal energy to kink-pair formation becomes relatively large compared with that of work done and hence the relative differences in the transition probability between various types of transitions from one site to the next become less as the temperature is increased.

b) Thermal Activation Analysis. The analysis of the activation parameters on the assumption of the Arrhenius rate equation has become popular since the early pioneering work by *Conrad* and coworkers in the 1960s [6.33]. However, it is only recently that detailed analyses over the whole temperature range have been successfully carried out for bcc metals based on experimental data down to helium temperature.

The plastic strain rate controlled by the Peierls mechanism is expressed, from (5.19), by

$$\dot{\gamma} = \dot{\gamma}_0 \exp\left(-\frac{E(\tau)}{k_B T}\right) , \tag{6.3}$$

for the smooth kink model. The pre-exponential factor $\dot{\gamma}_0$, which includes the frequency factor in the kink-pair formation, mobile dislocation density, etc., depends generally on stress and temperature to some degree. However, the kink-pair formation energy $E(\tau)$ and the temperature are related to $\dot{\gamma}$ as an exponential function and hence the change in $\dot{\gamma}_0$ is negligible compared with the change in the exponent. The derivative of the activation energy with respect to the stress, i.e.

$$v^* = -\frac{dE}{d\tau}, \tag{6.4}$$

has the dimensions of volume and is called the activation volume. From (6.3) we obtain

$$v^*(\tau) = k_{\mathrm{B}}T \left(\frac{\partial \ln \dot{\gamma}}{\partial \tau}\right)_T \tag{6.5}$$

and thus the activation volume can be obtained experimentally by the strain-rate dependence of the flow stress. The value of $E(\tau)$ can also be estimated from the activation volume and the temperature dependence of the flow stress as follows. Equation (6.3) can be regarded as approximately expressing the relation between three variables, $\dot{\gamma}$, τ, and T, and hence an equality $(\partial \ln \dot{\gamma}/\partial \tau)_T (\partial \tau/\partial T)_{\dot{\gamma}} (\partial T/\partial \ln \dot{\gamma})_\tau = -1$ holds. It follows then that

$$E(\tau) = -\left(\frac{\partial \ln \dot{\gamma}}{\partial (1/k_{\mathrm{B}}T)}\right)_\tau = k_{\mathrm{B}}T^2 \left(\frac{\partial \ln \dot{\gamma}}{\partial T}\right)_\tau$$

$$= -k_{\mathrm{B}}T^2 \left(\frac{\partial \ln \dot{\gamma}}{\partial \tau}\right)_T \left(\frac{\partial \tau}{\partial T}\right)_{\dot{\gamma}} = -Tv^* \left(\frac{\partial \tau}{\partial T}\right)_{\dot{\gamma}}. \tag{6.6}$$

In Fig. 6.15 we plot the value of τ at $\dot{\gamma} = 1.7 \times 10^{-4}\,\mathrm{s}^{-1}$ and that of strain-rate dependence as a function of temperature for high-purity iron single crystals with two different orientations [6.34]. Broken lines in the figure are for iron single crystals containing 150 at. ppm C. The values of activation volume and activation energy analyzed according to (6.5) and (6.6) by use of the experimental data of Fig. 6.15 are plotted as a function of stress in Fig. 6.16.

The $E(\tau)$ values obtained by (6.6) are based on the assumption that the $\dot{\gamma}_0$ value in (6.3) is insensitive to τ and T. The validity of this assumption can be simply checked by calculating the $E/k_{\mathrm{B}}T$ value for the stress giving a constant strain rate at various temperatures and by examining the constancy of this value. Figure 6.17 presents $E/k_{\mathrm{B}}T$ values at $\dot{\gamma} = 1.7 \times 10^{-4}\,\mathrm{s}^{-1}$ plotted against temperature, showing a constant value except below 30 K. The validity of the Arrhenius strain-rate equation of (6.3) has actually been confirmed for a variety of metals. The discrepancy at very low temperatures was first demonstrated experimentally for Ta [6.35] and later shown to be common to all bcc metals [6.36]. The failure of the Arrhenius rate equation near absolute zero is due to a quantum effect of dislocation vibration, i.e., the transition probability is finite even for the state where the potential cannot be overcome in the classical scheme. Except in this very low temperature region, $E \propto T$ for a constant strain rate., and hence if we replace the ordinate axis of the E-τ_y relation in Fig. 6.16 with temperature and exchange the ordinate and abscissa axes, the obtained curve should coincide with the τ_y-T curve in Fig. 6.15.

In the τ_y-T curve in Fig. 6.15, we see a hump at around 250 K. Corresponding to this hump, a peak appears in the v^*-T curve in Fig. 6.16. It has been shown that by alloying this hump in the τ_y-T curve is shifted to a lower temperature

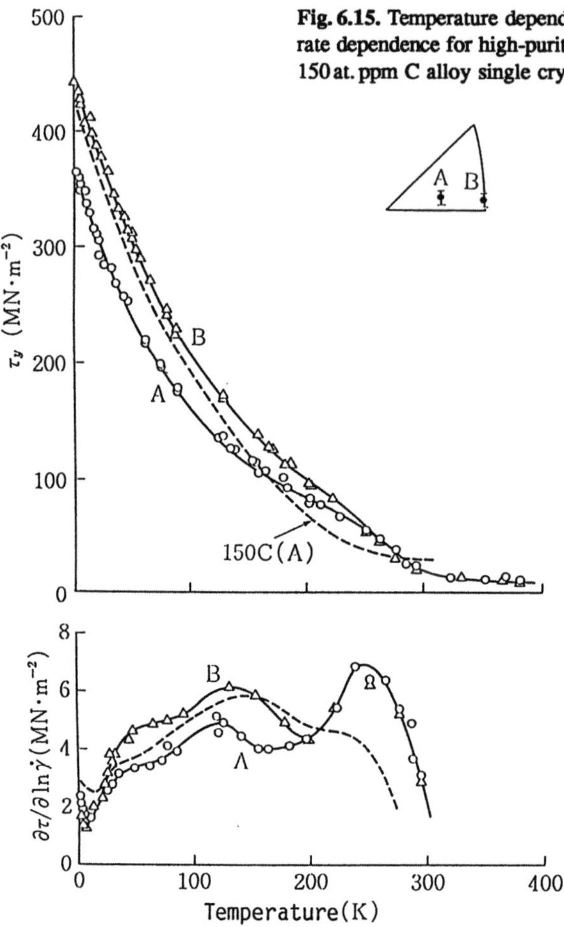

Fig. 6.15. Temperature dependence of the yield stress and its strain rate dependence for high-purity single crystals (solid lines) and Fe-150 at. ppm C alloy single crystals (broken lines) [6.34]

and finally disappears [6.37]. The mechanism of the appearance of this peculiar hump is interpreted to originate in a special shape of the Peierls potential for the {110}-type glide.[3] As mentioned in the previous section, the {110}-type glide takes place for an isotropic-type core. This type of core has a metastable configuration between two stable positions, and the Peierls potential has a camel-hump shape as illustrated in Fig. 6.18. For such a Peierls potential it is easy to show that the stress dependence of the kink-pair formation energy exhibits a hump, as shown in the figure, which is considered to correspond to the hump in the τ_y-T relation.

---

[3] Another interpretation of the hump in the τ_y-T curve is that the hump appears in the transition region of curves l_1 and l_2 in Fig. 5.5, i.e., the transition from the kink-kink interaction scheme to the line tension scheme of kink-pair formation [6.38]. However, this interpretation does not seem to be consistent with the facts that the hump appears only for the {110} glide, not for the {112} glide, and that the hump is shifted to higher stress by alloying.

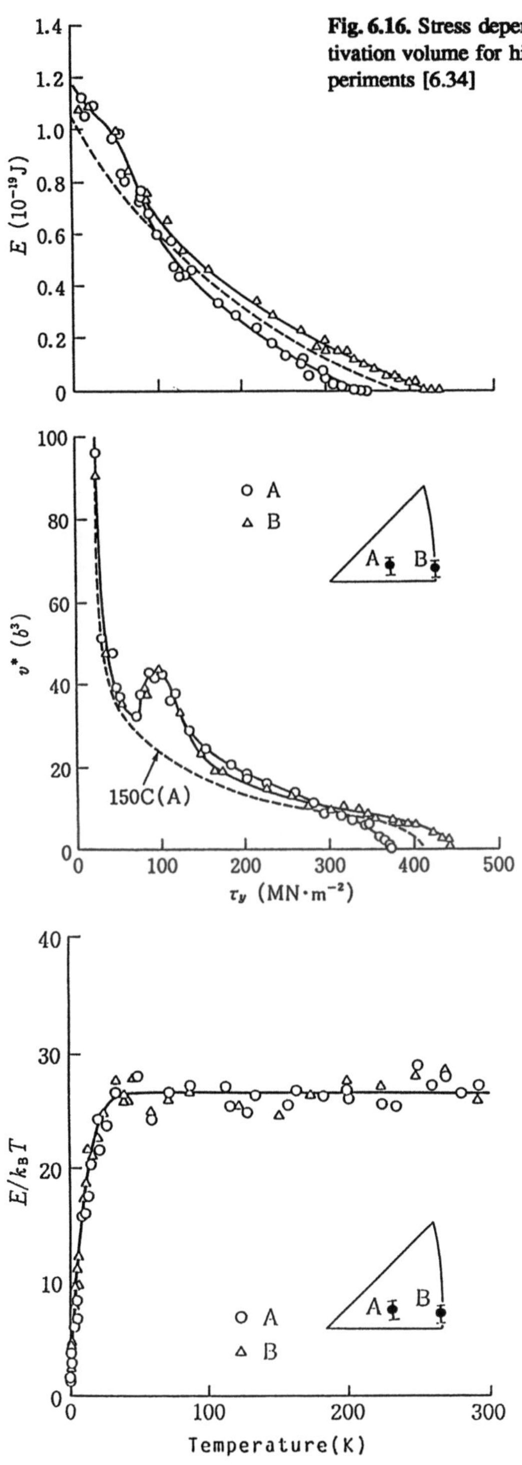

Fig. 6.16. Stress dependences of activation energy and activation volume for high-purity Fe obtained by tensile experiments [6.34]

Fig. 6.17. Temperature denpendence of $E/k_B T$ values for high-purity Fe at a constant strain rate, analyzed assuming an Arrhenius-type strain-rate equation (6.3) [6.34]

95

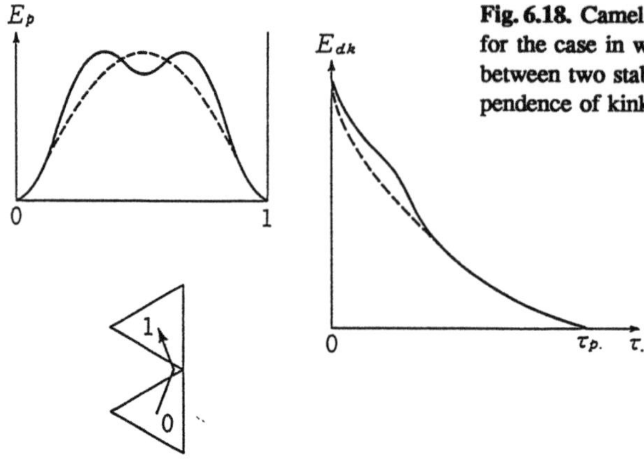

Fig. 6.18. Camel-hump Peierls potential (*upper left*) for the case in which there exists a metastable state between two stable positions 0 and 1, and stress dependence of kink-pair formation energy (*right*)

The shift of the hump to lower temperature and higher stress by alloying indicates that the metastable configuration becomes less and less stable on alloying. In addition to the disappearance of the hump, alloying causes the τ_y value in this temperature range to become lower than that of pure iron, as exemplified in Fig. 6.15 for Fe-150 at. ppm C by a dashed line. Such a solution-softening phenomenon is interpreted to be due to an enhancement of kink-pair formation by an interaction between solute atoms and the screw dislocation. Also, the shape of the Peierls potential is considered to be effectively changed from the camel-hump shape to a normal shape, shown by a dashed line in Fig. 6.18.

Detailed experiments for the whole temperature range and for the whole orientation range, as mentioned above for Fe, have been quite limited. Figure 6.19 shows the results for single crystals of an alkali bcc metal, potassium. The stresses are normalized with respect to that at 4.2 K. No hump is detected on the τ-T curve, probably because the dislocation in potassium is of polarized type and glides on {112} planes.

Among transition metals, we have so far presented the results only of Fe. The main reason for this is that in high-purity V and Nb, the so-called anomalous slip appears at low temperatures, which complicates the interpretation of temperature and orientation dependences. The "anomalous slip" is the {110}-type slip, however, not on the primary $(\bar{1}01)$ plane with the maximum shear stress but on the $(0\bar{1}1)$ plane, which is 60° off from the $(\bar{1}01)$ plane [6.40, 41], see Fig. 6.11. It has been revealed by electron microscopy that the glide directions are the coplanar [111] and $[\bar{1}11]$ [6.42]. The flow stress due to this anomalous slip is about half of the normal [111] slip in spite of the fact that the Schmid factor is only a half of the normal slip system.

The mechanism of the anomalous slip is interpreted in the following way [6.43, 44]. In high-purity bcc metals, the resistance to glide for an edge dislocation is small, and hence at first grown-in edge dislocations with various Burgers vectors glide at low stress to form long screw dislocations in various directions. If two produced screw dislocations with [111] and $[\bar{1}11]$ are close to each other, they

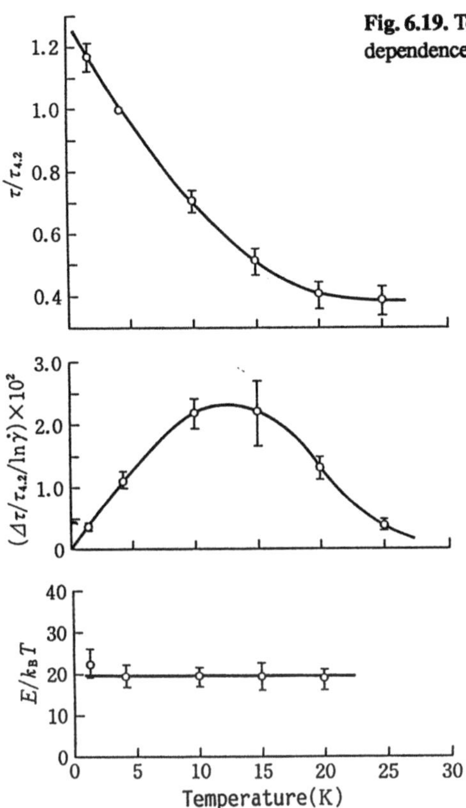

Fig. 6.19. Temperature dependences of flow stress, strain-rate dependence and $E/k_{\mathrm{B}}T$ for K single crystals [6.39]

exert a torque force on each other at their crossing point, which facilitates kink-pair formation at this position. Such a cooperative interaction between coplanar screw dislocations makes it possible to multiply dislocations at low stress, causing continuous plastic deformation. It is not yet clearly understood why the same phenomenon does not take place in Fe, but possibly grown-in edge dislocations in high-purity Fe are not so mobile as those in high-purity Va- and VIa-group bcc metals.

Another interesting role of the high mobility of edge dislocations in the plasticity is the so-called surface film softening effect [6.45, 46]. When the surface of a high-purity bcc metal is oxidized or another metal is deposited to form a thin surface layer, the flow stress decreases to about half at low temperatures. This phenomenon is interpreted as follows: At a stress much lower than the usual yield stress, cracks are formed at the surface layer, and due to a stress concentration at these cracks, multiplications of edge dislocations occur successively to produce macroscopic plastic strain, i.e., the plastic strain is brought about only by edge dislocations, the sources of which are the surfaces.

c) Plasticity of Solid-Solution Alloys. The effect of alloying on the plasticity of bcc metals controlled by the Peierls mechanism is not simple. Some common features of the bcc solid-solution alloys are: (a) a solid-solution hardening effect in the high temperature region which is almost proportional to the solute concentration; (b) a solid-solution softening effect at an intermediate low temperature region (an example being given in Fig. 6.15 for an Fe-C alloy).

The following are the three possible effects of alloying on the Peierls mechanism of screw dislocation. First, the kink-pair formation energy is partly supplied by the interaction between screw dislocation and solute atoms [6.47–49]. *Weertman* was the first to point out the possibility of solid-solution softening by the aid of the strain field of solute atoms in kink-pair formation [6.50]. Second, resistance to the kink migration process occurs due to an interaction between the kink and solute atoms. As a result, the dislocation velocity is determined not only by the kink-pair formation energy but also by the kink-solute interaction energy. The solid-solution hardening at high temperature is considered to be due to the resistance to kink migration. Third, the reduction of the kink velocity causes the formation of jogs on screw dislocations. In the regime of dislocation motion involving the kink–kink annihilation process mentioned in 5.2.3, the probability at higher temperature that two kinks formed on different planes will impinge to form a jog is considerable. The dragging force due to these jogs should also contribute to a part of the yield stress.

In high-concentration solid-solution alloys, resistance to the kink migration is induced by the self-energy variation of the kink due to statistical fluctuation of the solute concentration. A detailed theory by *Suzuki* [6.51] on this resistance successfully explains the experimental results for Fe alloys [6.52] and Ta alloys [6.53], except for a low temperature region. Later, *Suzuki* extended his theory to lower concnetration alloys by considering the interaction between a kink and individual solute atoms [6.54], and interpreted low-temperature experiments on Fe alloys by the Kyushu University group [6.26, 37, 55].

Generally, the low-temperature experiments near absolute zero on bcc solid-solution alloys are quite difficult, owing to twinning formation and brittle fracture. No systematic data at very low temperatures have yet been obtained except for limited experiments on Fe alloys by the Kyushu University group. In such a situation, it does not seem worth discussing the mechanism of the strength of bcc alloys in more detail.

7. Dislocation Motion in Semiconducting Crystals

Because the dislocation is an entity that carries plastic deformation, the physics of dislocations has been developed mainly for deformable metallic materials. In recent years, however, dislocations in semiconducting crystals, used for electronic devices, have also attracted considerable interest, because they sometimes play a crucial role in electronic properties: dislocations in semiconducting crystals act as electronically active centers just like impurities and point defects. On the other hand, motion of dislocations in these crystals is sensitively affected by the electronic properties of the crystal. As a result of interplay between dislocations and electrons, various interesting phenomena, not found in metallic crystals, are observed in semiconducting crystals. In this chapter, we review the characteristic behavior of dislocations in these crystals.

7.1 Introduction

The behavior of dislocations in semiconductors is basically governed by the structure and electronic properties of the dislocation core. Therefore, elucidation of the core state is essentially important in understanding dislocation behavior. High resolution electron microscopy has been a powerful tool in revealing the core structure to a considerable degree. Concerning the electronic state of the dislocation core, on the other hand, we have as yet not clear, unified view, mainly for the following two reasons: First, dislocations are usually introduced by plastic deformation in order to investigate their electronic properties, but in addition to dislocations various other point defects are simultaneously introduced and thus it is difficult to separate the contributions to the electrical properties due to different lattice defects. Second, the dislocation core may not have a unique state but is able to take different states.

Semiconducting crystals treated in this chapter are tetrahedrally coordinated as a result of sp^3 hybridization. They are group IV elemental crystals Si and Ge with diamond structure, III-V compounds mainly with zincblende structure and II-VI compounds with either zincblende or wurtzite structure. Covalent character decreases and ionicity increases, as a result of charge transfer, in the order group IV elemental crystals, III-V compounds and II-VI compounds. The tendency to wurtzite structure formation in II-VI compounds lies in the fact that the Madelung energy can be increased in this structure compared with the zincblende structure.

When we compare the Peierls stresses of III-V zincblende crystals and II-VI ones, the former are much larger than the latter, reflecting the difference in the strength of covalent bonding. This fact indicates that the Peierls potentials in these crystals are determined mainly by the bond breaking energy, unlike the case of metallic crystals. In this chapter, we first describe the present understanding of the atomic structure and the electronic structure of the dislocation core, and then present characteristic features of the dislocation mobility and the plasticity of covalent crystals.

7.2 Structure of Dislocations in Semiconducting Crystals

7.2.1 Atomic Structure

The diamond structure and zincblende structure are composed of two simple fcc sublattices, and $\{111\}$ planes stack as $a\alpha b\beta c\gamma a\alpha b\beta c\gamma\ldots$; if we consider the nearest atomic pairs as units then the stacking is the same as the fcc structure. In compound crystals, a, b, c layers and α, β, γ layers are composed of different atomic species. The wurtzite structure has a hcp structure with the stacking $a\alpha b\beta a\alpha b\beta\ldots$. The Burgers vector of active dislocations in the former type of crystal is $(1/2)\langle 110\rangle$ and that in the latter $(1/3)\langle 11\bar{2}0\rangle$; those are the same as fcc and hcp metals. The glide plane in the former crystals is $\{111\}$ as in fcc metals and that in the latter is not only the basal plane (0001) but also the prismatic plane $(1\bar{1}00)$.

Figure 7.1 shows atomic configuration of a 60° dislocation (defined later) in the diamond structure. Reflecting the polarity of the crystal in III-V and II-VI compounds, there are two types of edge dislocations. Depending on the species of atoms, group III (II) element or group V (VI) element, at the end of the extra half-plane, the edge dislocation is called an α-dislocation or β-dislocation,

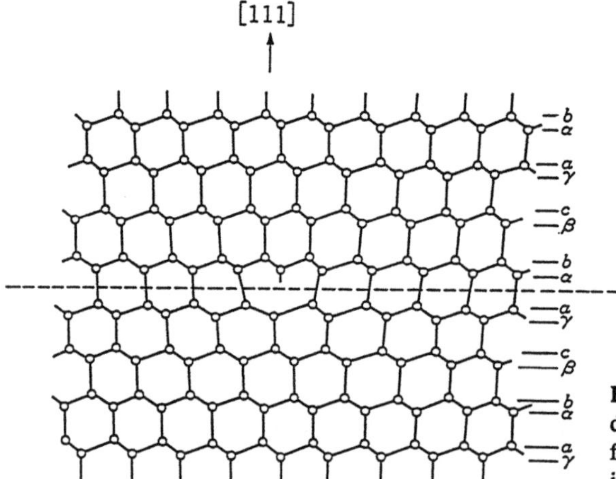

Fig. 7.1. Atomic configuration of a 60° dislocation in the diamond lattice viewed from the [$\bar{1}$01] direction. The broken line indicates a glide plane

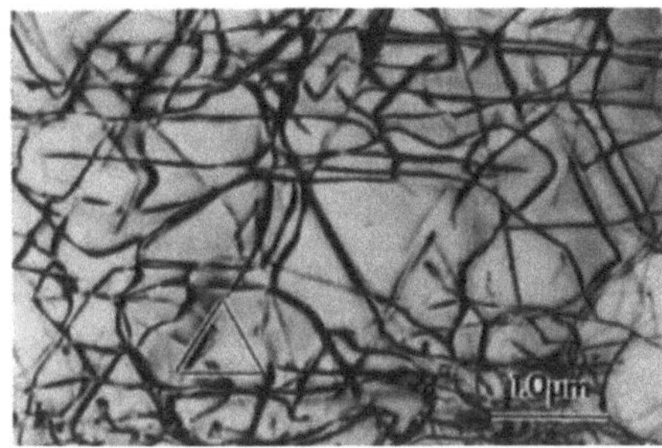

Fig. 7.2. Transmission electron micrograph showing glide dislocations in deformed Si. The foil has been cut parallel to the glide plane. The triangle in the figure indicates three $\langle 110 \rangle$ directions

respectively. The valleys of the Peierls potential in semiconducting crystals are oriented in close-packed directions, i.e., $\langle 110 \rangle$ or $\langle 11\bar{2}0 \rangle$, and hence the dislocation line tends to lie either in the direction parallel to the Burgers vector or making an angle of 60° to it. The former is the screw dislocation and the latter is called the 60° dislocation. Figure 7.2 shows a dislocation configuration in a deformed Si single crystal. It may be seen that the dislocations tend to lie along three Peierls potential valleys. In III-V and II-VI crystals, there are two types of 60° dislocations, the α-60° dislocation and the β-60° dislocation.

For many years, it was supposed that the 60° dislocation assumes the core configuration presented in Fig. 7.1. However, since the development, in the 1970s, of the weak-beam technique of electron microscopy, it has been revealed that the dislocations in almost all semiconducting crystals are dissociated into Shockley partials with a stacking fault in between them [7.1]. The stacking fault is of the intrinsic type as in fcc and hcp metals, the stacking fault plane being between the α layer and b layer (or equivalently between β and c, or γ and a). Before the dissociation was recognized, the glide plane of a dislocation had been considered to lie between a wide interlayer of $\{111\}$ stacking, i.e., a-α (or b-β or c-γ), but it is highly probable that the glide plane is in a narrow stacking layer of a α-b (or β-c or γ-a) in which the stacking fault exists. The dissociation scheme in which the glide plane coincides with the fault plane is called the glide set, while the dissociation structure which is produced from the undissociated configuration of Fig. 7.1 is called the shuffle set.

Figure 7.3 shows the atomic configuration of the glide set of the 60° dissociated dislocation. If we subtract or add one atomic row to the right side partial (90° partial) dislocation, then we obtain the shuffle set. There is no firm experimental evidence as to whether the real dissociated dislocation is of the glide set or of the shuffle set, but it is now widely believed that the dissociation scheme is the glide set. The reasons for this are that the self-energy of the glide set is considered to be lower than that of the shuffle set and that a recent high-resolution microscopy study on the core structure of the dissociated dislocation supported the possibility of the glide set [7.2].

In Fig. 7.3, open and filled circles are the same element for elemental group IV crystals but in III-V and II-VI crystals they correspond to different elements. We should note that if the dislocation is the glide set, the atomic species at

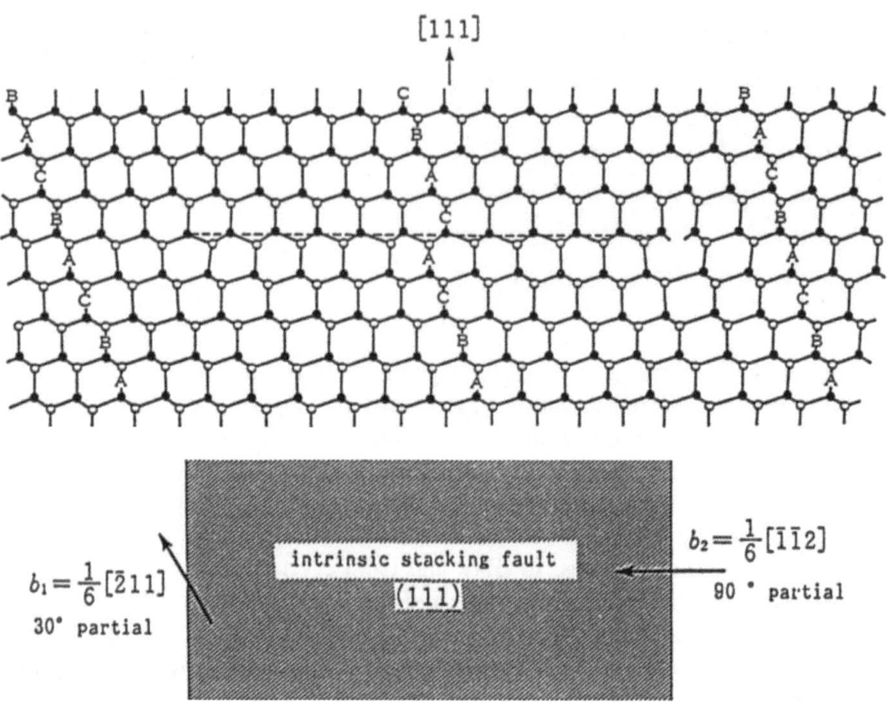

Fig. 7.3. (*top*) Atomic configuration of the dissociated 60° dislocation (glide set) in the diamond or zincblende lattice viewed from the dislocation line. Solid and open circles are for the same species of atoms in the diamond lattice and for different species in the zincblende lattice. (*bottom*) Dissociation of the dislocation viewed from the direction perpendicular to the glide plane

the end of the extra half-plane in α- and β-dislocations are opposite from those defined previously. Therefore, the definitions of α- and β-dislocations should be given not in terms of the atomic species at the end of the extra half-plane but in terms of the side of the extra half-plane with respect to the crystal polarity; i.e., in α- and β-dislocations the extra half-plane exists on the (111) side and $(\bar{1}\bar{1}\bar{1})$ side, respectively. The basal dislocations in the wurtzite crystal are dissociated in a similar manner; 60° dislocation is either the α-dislocation or the β-dislocation and the dissociation scheme is either the glide set or the shuffle set.

In Fig. 7.4 are shown various kinds of partial dislocations constituting 60° and screw dislocations in the diamond structure and the zincblende or wurtzite structure. In the diamond structure, there are two kinds of partial dislocations, the 30° and 90° partials. In the compound crystals, these two partial dislocations are further classified into α- and β-type since both of them have an edge component.

7.2.2 Electronic Structure of the Dislocation Core

As will be mentioned in the next subsection, the mobility of dislocations in semi-conducting crystals is affected sensitively by the electronic state of the crystal. This fact suggests that the electronic state of the dislocation core plays a signif-

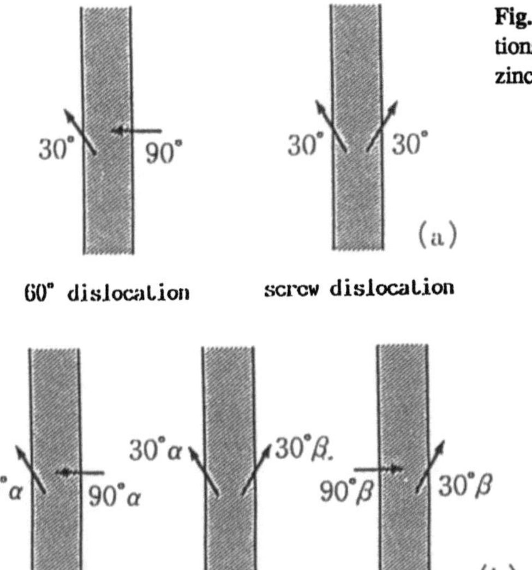

Fig. 7.4. Various types of dissociated dislocations in the diamond lattice (a) and in the zincblende or wurtzite lattice (b)

icant role in the dislocation mobility. In semiconducting crystals, each atom is more or less bound with four nearest neighbor atoms by covalent bonding. Consequently, at the center of an edge dislocation, there must be dangling bonds, having no partner atom to bind, along the dislocation line as shown in Fig. 7.1. Considering this fact, *Read* proposed in the 1950s the so-called *Read model*, where the dislocations in semiconducting crystals act as acceptors, and hence in n-type crystals, the dislocation captures free electrons around the dislocation line to form a cylindrical space-charge region [7.3]. More recently, the Read model was modified so that the wavefunctions of dangling electrons, aligned with a spacing of an atomic distance, overlap each other to form a one-dimensional band in the band gap, instead of localized centers. The one-dimensional band forms a half-filled band in the neutral state and thus acts as both acceptor and donor (electron trap and hole trap centers). This model was successfully applied to the effect of dislocations on the electrical properties of Ge [7.4].

In n-type crystals, the dislocation captures free electrons around it to form a positive space-charge region, as schematically shown in Fig. 5.a, while in p-type crystals the dislocation forms a negative space-charge region by capturing free holes as shown in Fig. 7.5b. For Ge, experimental results on the temperature dependence of the Hall coefficient in crystals containing various numbers of dislocations have been well explained on the assumption of the existence of half-filled dislocation bands [7.4]. However, experimental results from Si crystals cannot be interpreted fully by the above simple model; more than two dislocation bands must be assumed to explain the results [7.5, 6].

Actually, there are various kinds of partial dislocations, as mentioned in the previous subsection, and hence the electronic levels accompanying the disloca-

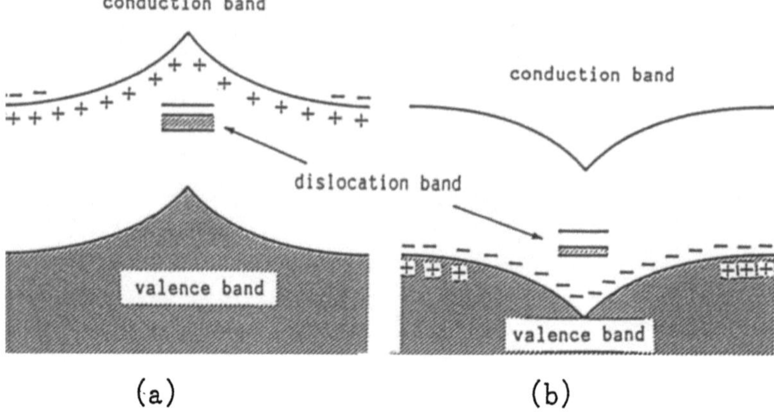

Fig. 7.5a,b. Electronic state around a dislocation in the half-filled, one-dimensional dislocation band model. (a) n-type crystal and (b) p-type crystal

tions cannot be simple. There are two dangling bonds per atomic distance for a 90° partial and one for a 30° partial, see Fig. 7.3, and hence the number of dangling bonds is increased to three times as many compared with the undissociated state. Trials have been made to detect these dangling bonds by ESR experiments [7.7–10]. Newly detected signals in Si after a plastic deformation have shown an anisotropy consistent with the Burgers vector of the introduced dislocations, confirming that the unpaired electrons detected are concerned with the dislocations.

Figure 7.6 shows atomic arrangements in the atomic layers adjacent to the glide plane viewed perpendicularly to the glide plane, (a) without reconstruction of dangling bonds and (b) with reconstruction. By the reconstruction where dangling bonds change their directions to form new bonds with each other as in (b), the net energy of the dislocation may be reduced at the expense of the bond bending strain energy. Such a reconstruction of dangling bonds is a well-known phenomenon in the surface state of semiconductors. The possibility of reconstruction in the dislocation cores has been pointed out by *Hirsch* [7.11], *Jones* [7.12], and *Marklund* [7.13], and is widely believed to hold for the dislocations in Si. The reasons are: (1) the intensity of ESR signal that is considered to be due to dangling electrons at the dislocation core is two orders of magnitude smaller than the estimated intensity on the assumption of three dangling bonds per atomic distance of the dislocation[1], and (2) theoretical calculations support the possibility of reconstruction [7.12, 13]. In compound semiconductors, however, the possibility of reconstruction is not known.

As a result of the reconstruction of the core of partial dislocations, a new type of point defect, the antiphase defect or the soliton defect, comes into existence. The antiphase defect is a dangling bond formed at the boundary between the two different phases of reconstruction as shown in Fig. 7.6b. It has been suggested

---

[1] Regarding this issue, no general agreement has yet been reached (see, e.g. [7.5]).

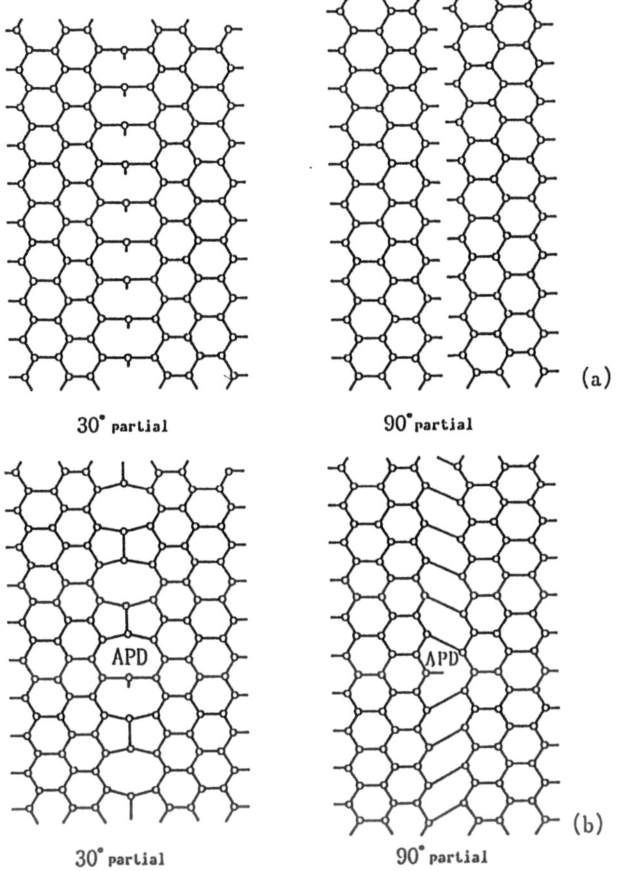

Fig. 7.6. Dangling-bond configurations in two atomic layers of the 30° partial and 90° partial viewed from the direction perpendicular to the glide plane: (a) for the unreconstructed state and (b) for the reconstructed state. APD indicates the antiphase defect which appears between different phases of reconstruction

that the ESR signal originates from the antiphase defect stabilized by combining with vacancies [7.15].

As mentioned so far, the old, simple picture of dislocations in semiconductors has had to be largely modified. The electronic state at the core of a neutral dislocation is a half-filled one-dimensional band in the band gap, as shown in Fig. 7.7, before reconstruction takes place, while after reconstruction the dislocation band splits into the filled bonding state and the empty antibonding state for the neutral state, as shown in Fig. 7.7b. The relative positions of these dislocation levels with respect to the valence and conduction bands must vary from crystal to crystal and also depend on the type of the partial dislocation. It may also happen that either of these levels is located in the valence or the conduction band. The electron occupation of these levels should change depending on the Fermi level; this fact is considered to be closely related to the dislocation mobility.

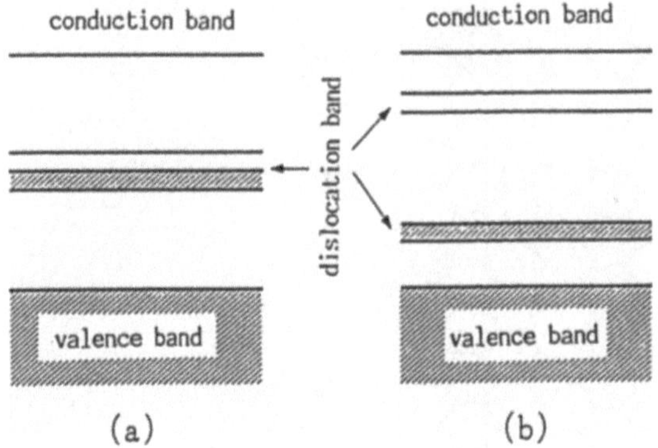

Fig. 7.7. Dislocation bands in the band gap: (a) before reconstruction and (b) after reconstruction

7.3 Mobility of Dislocations in Semiconducting Crystals

7.3.1 Experimental Facts

A number of experiments have been carried out on the dislocation mobility in Ge [7.14, 16–18] and Si [7.18–26] mainly by use of the etch-pit method. The measurements have been limited to a relatively high temperature range where the specimens are ductile. General features are:

(1) The dislocation velocity is expressed as a function of temperature and stress by

$$v = v_0 \left(\frac{\tau}{\tau_0}\right)^m \exp\left(-\frac{E}{k_{\mathrm{B}}T}\right) , \tag{7.1}$$

where v_0 and τ_0 are constants. The stress exponent m lies between 1 and 2, and the activation energy E is about 1.5 eV for Ge and about 2 eV for Si. Figure 7.8 shows the most reliable experimental data, obtained by the in situ X-ray topographic method, for the stress and temperature dependence of the dislocation velocity in high-purity Si crystals [7.26]. It had been reported in other experiments that the m value in (7.1) has a tendency to increase at lower stress, but *Imai* and *Sumino* clarified that this tendency is related to the pinning effect of the dislocation by impurities such as O and N and that the m value is unity for a wide stress range in high-purity Si [7.26].

(2) The dislocation velocity is dependent on the acceptor or donor concentration. Figure 7.9 illustrates the effect of doping on the dislocation velocity in Ge first reported by *Patel* and *Chaudhuri* [7.14]. A special feature is that, depending on the dopant species, a pronounced enhancement of the dislocation mobility can be produced, indicating a solid-solution softening effect rather than the usual solid-solution hardening effect. Similar phenomena have also been reported for Si; generally the higher the Fermi level the higher the mobility in Si and Ge [7.14, 21, 24, 25]. This mobility change is mainly due to the change in the E

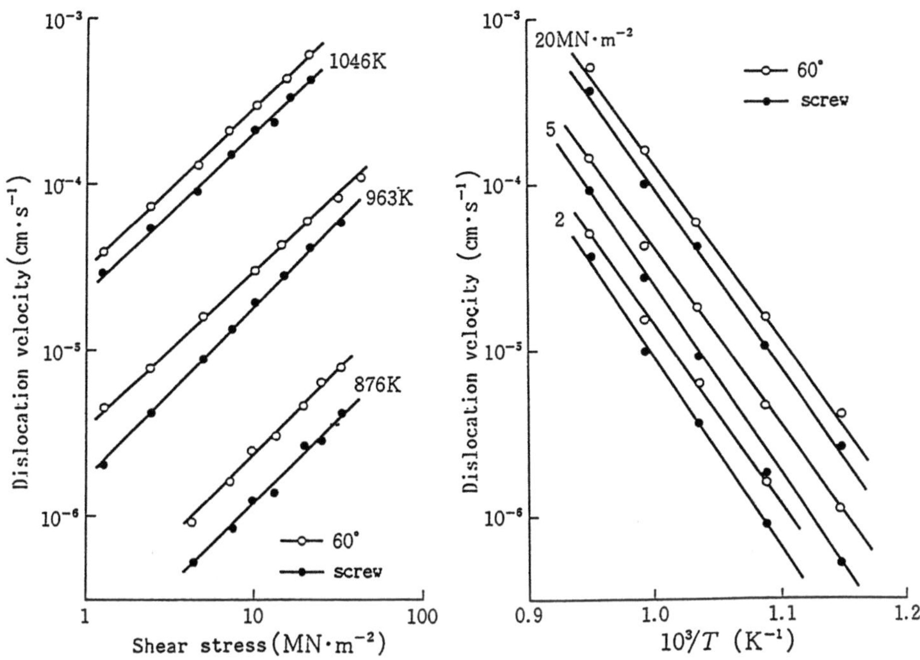

Fig. 7.8. Stress and temperature dependence of the dislocation velocity in high-purity Si measured by in situ X-ray topography [7.26]

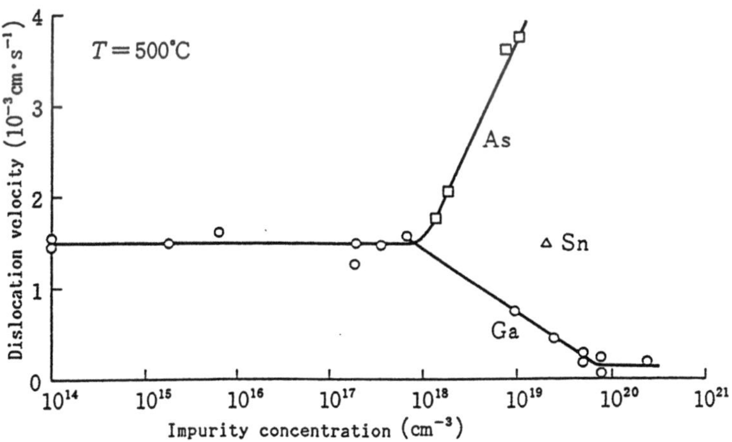

Fig. 7.9. Effect of doping on dislocation mobility in Ge [7.14]

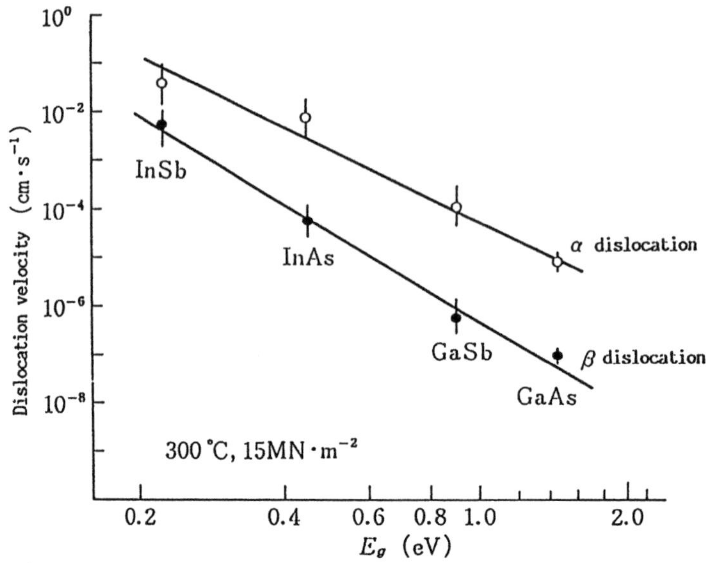

Fig. 7.10. Velocities of α- and β-dislocations in various III-V compounds plotted as a function of the band-gap energy at 0 K [7.35]

value in (7.1) due to doping but at the same time the prefactor is also reported to change [7.26].

(3) Within the same crystal, the dislocation velocity is dependent on the type of dislocation. The mobility of the 60° dislocation is different from that of a screw dislocation. Moreover, the mobility of the 60° dislocation depends on the direction of motion [7.27, 28] because the the dissociation of the 60° dislocation is not symmetric, see Fig. 7.4.

Equation 7.1 applies also to III-V compounds such as GaAs and InSb [7.29–35]. For III-V compounds it is known that the velocity of the α-dislocation is orders of magnitude higher than that of the β-dislocation. Figure 7.10 shows the dislocation velocities of α- and β-dislocations in undoped III-V compounds as a function of the band-gap energy E_g of the crystals, measured at 300°C under $\tau = 15\,\mathrm{MN/m^2}$ by *Choi* et al. [7.35]. The figure indicates that the dislocation velocity has a correlation with the band-gap energy and that the ratios of the velocities of α- and β-dislocations are almost constant.

Impurity doping effects have also been reported for III-V compounds [7.31, 32, 36]. The effect is particularly noticeable for the β-dislocation, but the dependence on the Fermi level is opposite to the case of Ge and Si. The mobility of the screw dislocation behaves similarly to the β-dislocation and hence it is considered that the 30° β-dislocation, which is the partial dislocation common to the screw and β-dislocations, governs the mobility of these two types of dislocations [7.36]. In zincblende crystals, the structure not only of the 60° dislocation but also the screw dislocation are not symmetric with respect to the direction of motion (see Fig. 7.4), and hence the α-dislocation, β-dislocation and screw dislocations should all have different mobilities with respect to the direction of their motion. No such distinction, however, has been reported in any experiment.

For II-VI crystals, attempts to measure the dislocation mobility by the etch-pit method have not been successful, because dislocations do not glide continuously over a long distance. The yield stresses of II-VI compounds are much lower than those of group IV crystals and III-V compounds and of the same order as those of metals. The temperature dependence is weak at room temperature. Namely, the long-range motion of dislocations, which governs the macroscopic deformation, is controlled not by the Peierls mechanism but predominantly by other obstacles. In situ electron microscope observation indicates, however, that the short-range motion of dislocations in II-VI compounds is controlled by the Peierls mechanism even at room temperature [7.37], and thus mobility measurements may be possible at lower temperatures.

7.3.2 The Mechanism Controlling the Mobility

The motion of dislocations in highly covalent crystals such as group IV and III-V semiconductors is considered to be treated by the abrupt kink model described in Sect. 5.2.3. However, since the dislocations in semiconducting crystals are dissociated into Shockley partials, the situation is not so simple as an isolated dislocation; two partial dislocations may undergo correlated motions elastically interacting with each other. A theory of the Peierls mechanism of the dissociated dislocation shows that below a critical stress τ_c the kink-pair formations of two partials are related to ech other, giving complicated results, but above τ_c the result is of the same form as (7.1), i.e., $m = 1$ [7.38]. The results of the dislocation mobility in high-purity Si in Fig. 7.7 are consistent with the theoretical result, (5.24) or (5.34). Depending on the relative magnitude of the kink collision length L (5.31) and the mean dislocation length or the mean separation of strong obstacles Λ, the dislocation velocity equation is either (5.32) or (5.33). To estimate the L value experimentally, one needs to measure the dependence of the dislocation segment length on the velocity. Such an experiment has actually been performed for Si by electron microscopy [7.39–41].

In semiconducting crystals, the activation energy of the dislocation velocity involves both the kink-pair formation energy E_{kp} and the kink migration energy E_p^k. This situation is different from bcc metals, where the activation energy contains only the kink-pair formation energy. However, the ratios of the two contributions, i.e., E_{kp} to E_p^k, in the experimentally obtained activation energies in semiconductors are not known.

Not only E_p^k but also E_{kp} is considered to be affected by the electronic state of the kink, but no reliable theoretical calculation has yet been performed for this effect. The doping effects on the dislocation velocity must be a result of the change in E_p^k and/or E_{kp} with the change in the electronic state of the dislocation. Various mechanisms have been proposed for the doping effects: a change of E_{kp} due to a change of the electronic charge of the dislocation (the larger the charge the greater the reduction of the effective line tension of the dislocation, which

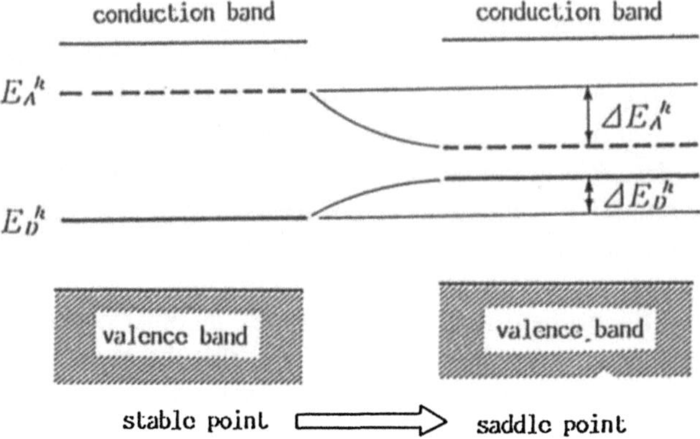

Fig. 7.11. Donor level E_D^k and acceptor level E_A^k accompanying a kink at the stable point and the saddle point

results in a decrease of E_{kp} [7.42]); a change of the thermal equilibrium kink concentration due to a change in the Fermi level [7.43]; and a change in E_p^k due to a change in the electronic state of kinks depending on the Fermi level [7.44]. The first effect is considered to be insufficient to explain the experimental results, while the second and the third effects play major roles; however, it is not yet known which effect is dominant.

Consider the case where there exist a donor level E_D^k and an acceptor level E_A^k in the band gap accompanying kinks, as shown in Fig. 7.11. In the neutral state the level E_D^k is filled and E_A^k is empty. In the thermal-equilibrium state, the ratio of the concentrations of negatively or positively charged kinks to that of the neutral kinks are expressed, respectively, as

$$\frac{c_k^-}{c_k^0} = \exp\frac{(\mu_f - E_A^k - eV)}{k_B T} ,$$
$$\frac{c_k^+}{c_k^0} = \exp\frac{(E_D^k - \mu_f + eV)}{k_B T} ,$$

(7.2)

by taking account of the contribution of an electronic energy to E_k of (5.20) [7.43]. Here, μ_f is the Fermi energy and the term eV signifies the electrostatic potential due to the charge.

At high temperature and low stress, where the dislocation velocity is represented by (5.24), the dislocation velocity is composed of three terms, v_0, v^+ and v^-, from three kinds of kinks: neutral, positively charged and negatively charged, i.e.

$$v = v^0 + v^- + v^+ = v^0 \left[1 + \exp\left(\frac{\mu_f - E_A^k - eV + \Delta E_p^{k(-)}}{k_B T}\right) \right.$$
$$\left. + \exp\left(\frac{E_D^k - \mu_f + eV + \Delta E_p^{k(+)}}{k_B T}\right) \right] .$$

(7.3)

Here, $\Delta E_p^{k(+)}$ is the change in the kink migration energy due to the electronic charge. In the course of the kink migration surmounting the Peierls potential of the second kind, the levels E_A^k and E_D^k should vary. At the saddle point of the kink migration, the difference in energy of the bonding state and the anti-bonding state at the kink is expected to be reduced and hence the E_A^k and E_D^k levels will change as shown in Fig. 7.11. As a result, when the kink is negatively charged the kink migration energy is reduced by ΔE_A^k compared with the neutral kink and when positively charged it is also reduced by ΔE_D^k because of the absence of an electron in the level E_D^k. Of course, if the kink level E_D^k or E_A^k is located in the valence or the conduction band, no such effect is expected. In summary, $\Delta E_p^{k(+)}$ is the sum of the change in the Peierls potential due to an electrostatic relaxation and that due to the above-mentioned electronic energy.

Thus, the change in the Fermi level induces a change in the dislocation velocity by means of two effects: the change of the thermal-equilibrium concentration of kinks and the modification of the Peierls potential for the kink migration. It is not known, however, which of the two effects is dominant.

The α- and β-dislocations in III-V compounds must have quite different levels of E_A^k and E_D^k from each other, because the atomic species which produces dangling bonds are different between the two dislocations. Furthermore, due to the ionic nature of the lattice the α- and β-dislocations should be differently charged even if they do not capture carriers. These circumstances will bring about the difference in mobility and the different effects of doping for the α- and β-dislocations [7.45].

7.4 Effect of Electronic Excitation on the Dislocation Mobility

Owing to the fact that the dislocation velocity in semiconducting crystals is sensitively dependent on the electronic state of the crystal, it may also be affected by electronic excitation of the crystal by light illumination or electron irradiation, or by injection of carriers in a dioide crystal. Actually, it was reported earlier that the hardness of semiconducting crystals is influenced by illumination with light [7.46], and later the so-called photoplastic effect was discovered for II-VI compounds. More recently, it has been reported that during operation of a light emitting diode of a III-V compound, an anomalous multiplication of dislocations takes place and deteriorates the diode. The dislocation multiplication during operation of the device is mostly due to climb motion of dislocations, which is interpreted as being induced by absorption of excess point defects existing in the crystal [7.47]. Also, glide motion of dislocations in optoelectronic devices has been reported to be enhanced by carrier injection [7.48–51]. In recent years, a quantitative evaluation of this effect has been carried out in well-defined, controlled conditions initially for n-GaAs [7.52], and later not only for various III-V compounds [7.53, 54] but also for Si [7.55]. So far, the quantitative measurements have been performed by electron-beam excitation for III-V compounds

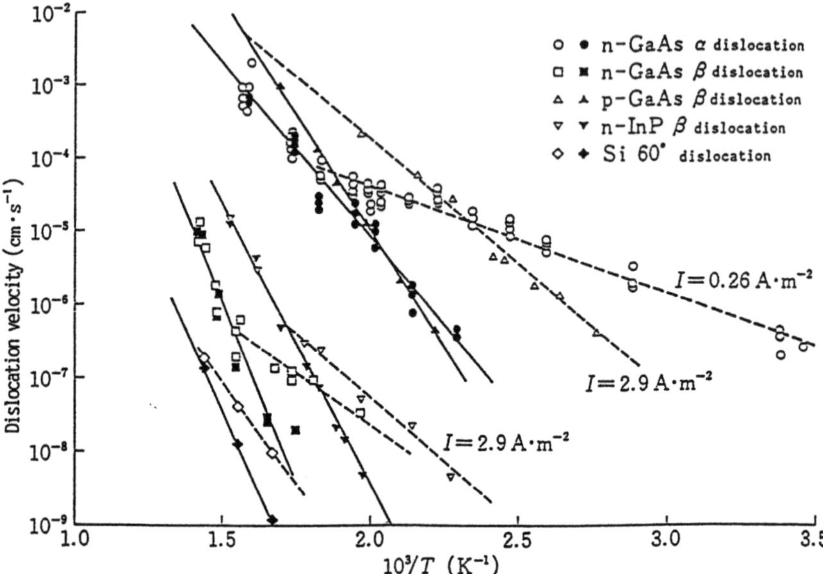

Fig. 7.12. Effect of crystal excitation on the dislocation velocity in various crystals. Open marks for the data under electron beam excitation (for III-V compounds) or under laser light excitation (for Si), and filled marks for the data in the dark. I is the electron beam current density. Stress is $294\,\mathrm{MN/m^2}$ for Si and $26\,\mathrm{MN/m^2}$ for the other samples

and by laser light excitation for Si, but both sets of data show essentially the same results.

Examples of the effects of excitation on the dislocation mobility in various crystals are shown in Fig. 7.12. Filled symbols in the figure indicate the data in the dark, i.e., without excitation, and open marks under excitation. The dislocation mobility in the dark obeys a single Arrhenius relation with a single activation energy given by (7.1), but under excitation the mobility follows two different activation processes depending on temperature. In the high temperature range the mobility is the same for both states and in the low temperature range the mobility is much enhanced by excitation. In summary, the dislocation velocity can be represented by the sum of two terms; one is the usual thermal-activation term (7.1) and the other is a new activation term with a reduced activation energy which appears under excitation, i.e.,

$$v_{\mathrm{excit}} = v_1 \exp\left(-\frac{E_\mathrm{T}}{k_\mathrm{B}T}\right) + v_2 \exp\left(-\frac{E_\mathrm{T} - \Delta E}{k_\mathrm{B}T}\right) . \tag{7.4}$$

Here, E_T is the activation energy in the usual thermal-activation process and ΔE is the reduction of the activation energy by excitation. The excitation effect, of course, depends on the excitation intensity. The prefactor v_2 in the second term is found to be almost proportional to the intensity of the electron beam used for excitation [7.53].

Because the excitation or carrier injection causes the shift of the pseudo-Fermi-level of the crystal, the above effect may have a similar origin to the doping effect mentioned previously. However, the change of the activation energy in the excitation effect is incomparably larger than that in the doping effect, and hence the origin of the excitation effect cannot be the same as that of the doping effect. The most plausible mechanism of the excitation effect is a nonradiative recombination enhancement mentioned below.

In semiconducting crystals not only dislocations but lattice defects in general form deep levels in the band gap, which act as trapping centers for excess carriers and as preferential recombination centers for electron-hole pairs. Furthermore, the strain field around a defect is modified by the electronic state of the defect, particularly for compound semiconductors with high ionicity, and the recombination process generally involves a change in the strain energy. As a result, the energy released upon electron-hole recombination at the defect is very often converted to lattice phonons, instead of photon emission.

It is known that dislocations in semiconducting crystals actually act as strong nonradiative recombination centers. The effect of the phonons emitted upon the nonradiative recombination process at a defect – on an enhancement of the migration of the defect itself or on the creation of a new defect – is called the *recombination enhanced defect reaction*, abbreviated REDR. For example, an anomalous enhancement of the recovery process of excess point defects in III-V compounds by carrier injection has been interpreted as REDR [7.56, 57]. The theoretical treatment of the REDR was first given by *Weeks* et al. [7.58] for a simplified model and then in more detail by *Sumi* [7.59]. The results show that for the case where the reaction efficiency is high REDR occurs at the rate following the relation $\exp[-(E_T - E_{ph})/k_B T]$, where E_{ph} is the energy emitted at the nonradiative recombination process (either in the capture process of minority carriers or in recombination with another carrier), and $E_T(> E_{ph})$ is the usual activation energy. This result corresponds exactly to the second term of (7.4).

The microscopic process where REDR takes place in the Peierls mechanism of the abrupt kink is either the enhancement of the kink migration and/or the enhancement of the initial unit kink-pair formation process, see Fig. 5.7. In the case where the dislocation velocity is expressed by (5.33), the enhancement only of the kink migration process does not produce an increase of the dislocation velocity, and hence enhancement of the initial unit kink-pair formation has to take place. However, detailed elementary mechanisms have not yet been clarified. The prefactor v_2 includes excess minority carrier concentration, carrier diffusion rate and the cross section for trapping by the dislocation. As seen in Fig. 7.12, the v_2 values, obtained by extrapolation of the velocity data of the low temperature region to $1/T = 0$, vary greatly for different types of dislocation in GaAs. Such differences are interpreted to originate in the difference in the capture cross section of the carrier for different types of dislocations due to the different charge states of the dislocations [7.60].

It is highly probable that the enhancement of the dislocation mobility by the nonradiative recombination is a phenomenon common to all semiconducting crystals, and it is expected that quantitative measurements of the effect will be performed in the future for various crystals. These results would give us information on the dislocation level or on the charge state of the dislocation.

7.5 Photoplastic Effect in II-VI Compounds

7.5.1 Plasticity and Dislocation Motion in II-VI Compounds

Before approaching the photoplastic effect, we have to understand the difference between II-VI comounds and other semiconducting crystals with the diamond structure. As decribed in Sect. 7.3.1, we cannot measure the dislocation velocity by the etch-pit method for II-VI compounds at room temperature. The reason for this is that dislocation multiplied in the first loading mostly do not move in the second loading but instead new multiplications occur in other places [7.61]. However, multiplied dislocations in group IV and III-V compound semiconductors usually keep on moving continuously for repeated loadings, which makes it possible to measure the dislocation mobility accurately, as already presented previously. In the latter case, the macroscopic strain rate can be expressed by $\dot\gamma = \varrho b\bar{v}$, (5.19), using the moving dislocation density ϱ and the average dislocation velocity $\bar{v}$ under an applied stress. For II-VI compounds, however, it is more appropriate to express the macroscopic strain rate by

$$\dot\gamma = \dot{n}b\bar{l} \tag{7.5}$$

using the multiplication rate $\dot{n}$ (total length of the dislocation generated in unit volume in unit time) and the mean glide distance of dislocations $\bar{l}$ (average value of the distances travelled by dislocations). In direct in situ observation of deformation in II-VI compounds, the glide of dislocations is relatively smooth and continuous, indicating that the microscopic deformation mechanism is the Peierls one, as mentioned previously. The multiplication process in these crystals shows, however, a characteristic feature. A series of micrographs showing one generation cycle of a dislocation multiplication from a source in a CdTe crystal is presented in Fig. 7.13. It takes quite a long time for the source dislocation to turn around a pole P in the figure, while the emitted dislocations glide with a very high velocity so that their motion can hardly be recognized. The slow motion of the source dislocation is due to the much reduced effective stress on it owing to the back stress of the line tension of the bow-out segment.

Let the length of the Frank-Read source in a bulk crystal be λ, then the source cannot be operated below a critical stress $\tau_c = \mu b/\lambda$. In the case where the velocity of free motion of straight dislocations at a stress $\tau \sim \tau_c$ is high enough, the macroscopic deformation is controlled by the multiplication process. On the other hand, if the dislocation velocity does not become high enough to yield the usual strain rate unless $\tau \gg \tau_c$ is satisfied, the macroscopic deformation is controlled by the average velocity of the multiplied dislocations. In the former

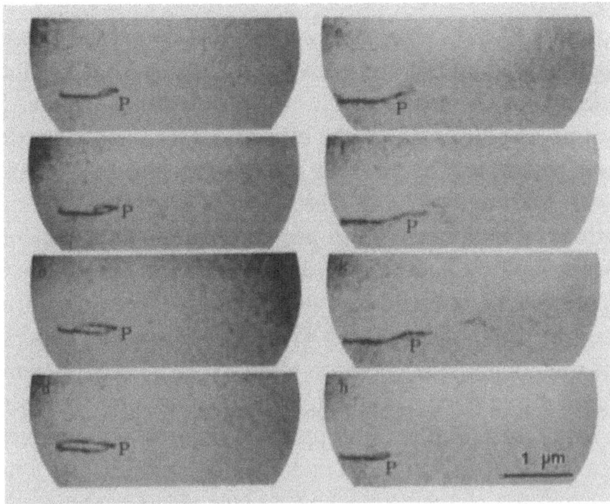

Fig. 7.13a–h. Electron micrographs of an example of the dislocation multiplication process observed in CdTe by an in situ deformation experiment [7.61]. (From left top to right bottom is shown one cycle of dislocation emission from a pole P)

case, the strain rate is expressed by (7.5) and in latter case by (5.19). The τ_c value will have a similar order of magnitude for different crystals, while the mobility of the dislocation varies by many orders of magnitude from crystal to crystal. Thus, the distinction between the above two types of expressions for the strain rate is determined by the dislocation mobility. In such crystals as in II-VI compounds, where the Peierls stress is relatively low, the deformation is usually multiplication controlled, at least at room temperature. Even in group IV semiconductors and III-V compounds, when the temperature is high enough the situation will become similar to II-VI compounds mentioned above.

Since $\dot{n}$ in (7.5) in II-VI compounds is controlled by the velocity of the source dislocation, which is controlled by the Peierls mechanism and obeys an Arrhenius-type equation, the strain rate is expressed by

$$\dot{\gamma} = \dot{n}_0 b \bar{l} \exp\left(-\frac{E}{k_{\mathrm{B}} T}\right) . \tag{7.6}$$

Here, $\dot{n}_0$ is a prefactor including the source density, and the activation energy E is the kink-pair formation energy as a function of the effective stress acting on the source dislocation.

7.5.2 Photoplastic Effect

The photoplastic effect is the name for the effect of light illumination on the plasticity of crystals. In the 1960s, an almost reversible change of the flow stress was reported for alkali halide crystals containing F-centers, on illumination with light which excites the F-centers [7.62]. This phenomenon has been interpreted as due to a change in the electrostatic interaction between the dislocation and the F-center by electronic excitation of the center.

For semiconducting crystals, experiments on the effect of light illumiunation on their hardness have been performed by several groups, but the results are not necessarily consistent with each other [7.63]. The effect of crystal excitation on the mobility of dislocations, mentioned in the previous subsection, can be regarded as a kind of a photoplastic effect. Before their effect was established, however, a photoplastic effect was discovered in 1968 for CdS, where the flow stress is drastically increased by light illumination [7.64]. More recent experiments have clarified that this phenomenon is common to all the II-VI compounds. An example of the stress-strain curve for a CdTe single crystal showing the photoplastic effect is given in Fig. 7.14.[2] Figure 7.15 presents data on temperature dependence of the yield stress both in the dark and under illumination for various II-VI compounds [7.66].

Features of the photoplastic effect in II-VI compounds are:

1) The effect is maximum for light whose wavelength is slightly longer than the absorption edge.
2) The effect increases with increasing light intensity and is saturated at relatively low light intensity, of the order of 10^3 lux.
3) The lower the temperature, the larger the effect, see Fig. 7.15.

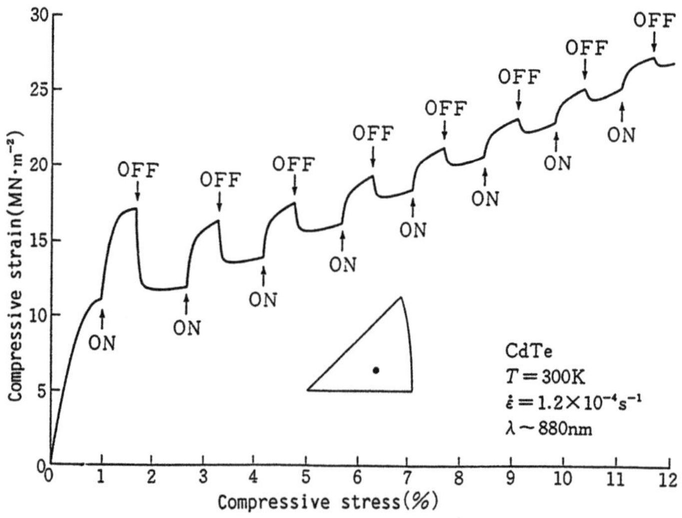

Fig. 7.14. An example of a stress-strain curve in a CdTe single crystal showing the photoplastic effect. The crystal has been illuminated intermittently (between ON and OFF). Experimental conditions are given in the figure

---

[2] A few experiments have been reported where illumination causes a softening of the crystal [7.65], but here we treat only a noticeable hardening effect.

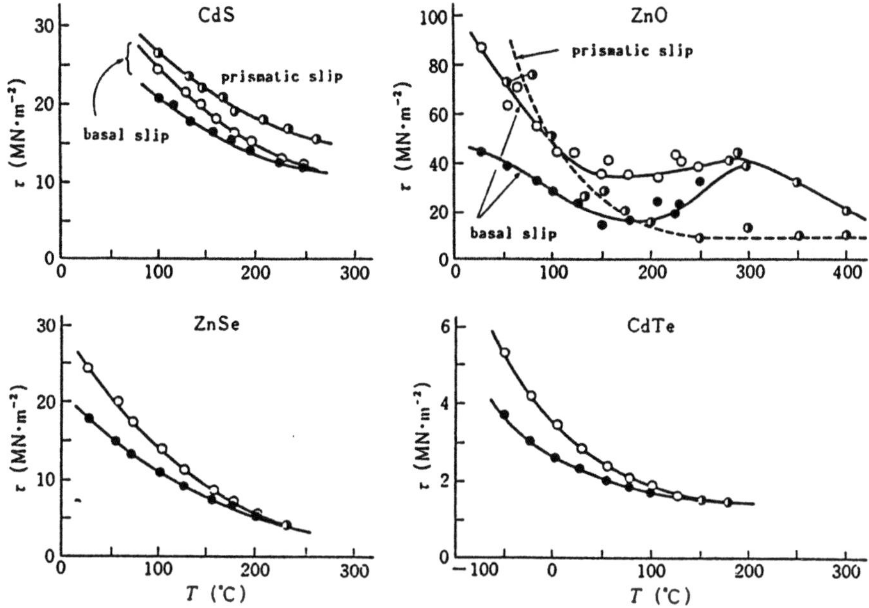

Fig. 7.15. Temperature dependence of yield stress in the dark (●), and under illumination (o) for four II-VI compounds (CdS [7.67], ZnO [7.68], ZnSe [7.69], CdTe [7.70]). o indicates no effect of illumination

4) In wurtzite crystals, the effect is not observed for prismatic slip, see Fig. 7.15.
5) The effect decreases as the strain increases, see Fig. 7.14.
6) The work-hardening rate and the rate of accumulation of dislocations with strain are larger for deformation under illumination than in the dark.
7) The so-called infrared quenching effect is observed, where a simultaneous illumination with an infrared light and the interband excitation light drastically reduces the photoplastic effect.
8) The photoplastic effect exhibits the same wavelength dependence as the dislocation charge [7.69]. The dislocation charge in II-VI compounds has been estimated by direct measurements of the electric charge which glide dislocations carry during deformation [7.65, 72].

What is the mechanism for the photoplastic effect? There are two possible effects of light illumination on the dislocation motion: one is a change in the dislocation mobility through a change in E in (7.6), and the other is a change in the mean free path l of multiplied dislocations. The crystal excitation effect for III-V compounds and Si mentioned in the previous section, corresponds to the former effect, but there, the effect is an enhancement, whereas in the photoplastic effect in II-VI compounds the effect must be opposite, i.e., a reduction of the mobility. The mechanism of the photoplastic effect based on the mobility change has been proposed by *Osip'yan* and *Petrenko* [7.69]. They postulated that the dislocation

117

mobility is controlled by a kind of Peierls potential originating in an electrostatic interaction between a charged dislocation and the periodic ions of the lattice; the electric charge of the dislocations is determined by the dynamic equilibrium state of the charges between the electron-trapping centers and the moving dislocations. They thought that the illumination causes an increase in the dislocation charge, which increases the Peierls potentials of electrostatic origin and reduces the dislocation mobility. In order to observe directly the effect of illumination on the dislocation velocity, *Nakagawa* et al. performed in situ transmission electron microscope observations for basal glide dislocations in CdS [7.73]. They introduced an optical fibre inside an electron microscope to illuminate the tensile specimen with laser light, and observed the effect of illumination by laser light on the dislocation behavior. In this experiment, since electron beam irradiation can cause the same effect as the illumination with light, the electron beam used to observe the positions of dislocations was emitted stroboscopically for short periods of time and thus the dislocation velocities both in the dark and under laser illumination were estimated without electron beam irradiation. In Fig. 7.16, velocities of individual dislocations in the dark and under illumination are plotted. No appreciable difference in the velocity between in the dark and in the light is recognized. This result does not support the interpretation that the origin of the photoplastic effect lies in the change of the dislocation mobility. On the other hand, if we assume that the mean free path of dislocations is decreased by illumination, we expect an increase in the accumulation rate of the dislocation and hence an increase in the work-hardening rate. In fact, the magnitude of the photoplastic effect and the results mentioned in (6) have been shown to be consistent with each other [7.66, 73]. Consequently, we may conclude that the reduction of $\bar{l}$ by illumination is the cause of the photoplastic effect.

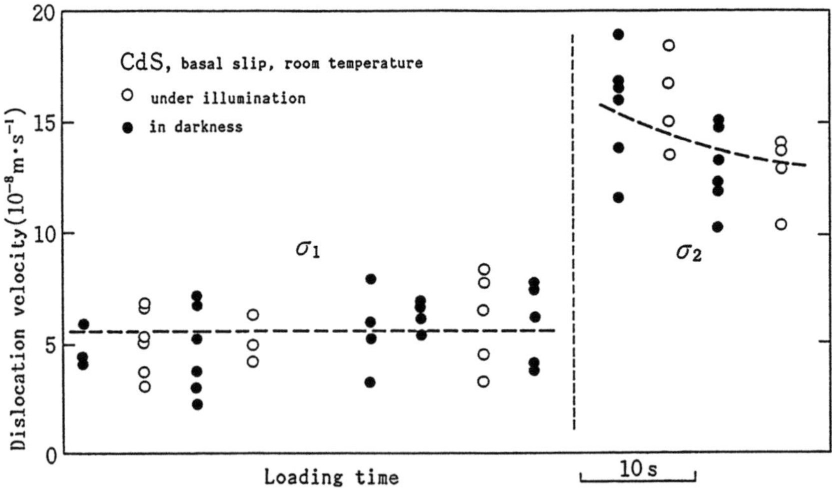

Fig. 7.16. Velocities of individual dislocations in CdS under illumination by laser light (o) and in the dark (●) as measured by an in situ tensile experiment in a transmission electron microscope

However, why the $\bar{l}$ value is decreased by illumination is not clearly understood yet. According to the in situ transmision electron microscope observations, the screw dislocations have been shown to have the lowest velocity and to undergo frequent cross-slip [7.37]. As a result of frequent cross-slip, the jog density on the screw dislocation increases with increasing distance of the dislocation glide and eventually the screw dislocation becomes immobile due to the increased jog dragging force. Thus, the frequency of jog formation seems to be a dominant factor in determining the $\bar{l}$ value. From the fact that illumination causes an increase of the dislocation charge which is closely related to the photoplastic effect as mentioned before, jog formation is possibly controlled by an electronic interaction between the charged dislocation and charged point defects. The dislocation on prismatic planes in wurtzite crystals are considered to be uncharged and hence no photoplastic effect is expected. In any case, compared with other semiconducting crystals, II-VI crystals so far used for experiments have been of much lower quality, with higher concentrations of defects. In this respect, the production of well-controlled, high-quality II-VI crystals is required in order to elucidate the interaction between dislocations and point defects.

For more details of dislocations in II-VI compounds, the reader is referred to a review paper on this subject [7.74].

8. High-Temperature Deformation of Metals and Alloys

The special aspect of high-temperature deformation is that the effect of diffusion is pronounced. The dislocations can climb as well as glide so that recovery due to dislocation rearrangement and pair annihilation proceeds quickly. Sometimes recrystallization occurs during deformation. For polycrystalline metals, grain boundary sliding and migration also play an important role. Under a very low stress, the deformation occurs not by dislocation motion, but exclusively by diffusion. In this chapter, the characteristics of high-temperature deformation are shown by using the deformation mechanism map. The problems special to high-temperature deformation, and the experimental techniques to settle them, are given to help the understanding of the deformation mechanism described in the following chapters.

8.1 Deformation Mechanism Map

The situation of high-temperature deformation is easily seen from the deformation mechanism map. The conditions necessary for a particular deformation mechanism to dominate are given in the map as a region in stress–temperature space.

This way of expression was first proposed by *J. Weertman* and *J.R. Weertman* [8.1, 2] in 1965 and studied in detail by *Ashby* [8.3] in 1972. Since then, the practical usefulness of the map has been widely recognized and many investigations have been conducted. Figure 8.1 is a schematic representation of the map drawn by *Frost* and *Ashby* [8.4].

Since the flow stress σ is proportional to the shear modulus μ for most of the deformation mechanisms, the stress axis is normalized by μ in order to remove the differences between the materials. The temperature axis is also normalized by the melting point T_m (absolute temperature), i.e., the homologous temperature is used for generalization. In this representation, the map is almost independent of the materials concerned as long as their bonding nature, such as metallic or covalent, and their crystal structure are the same. In the following, the essentials of each deformation mechanism will be described together with the corresponding problems with the map and improvements that have been made since the work of *Ashby*.

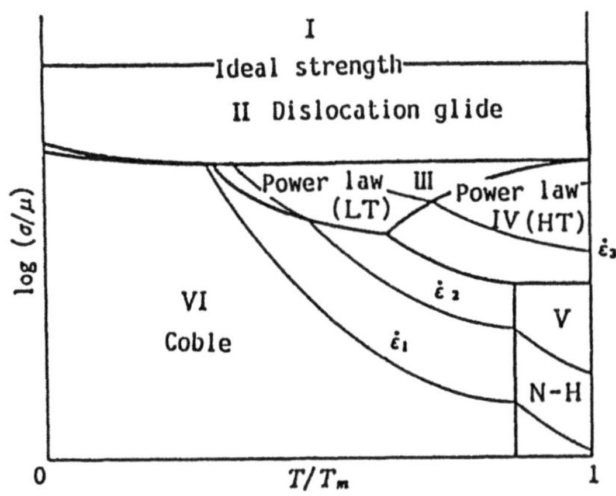

Fig. 8.1. Schematic representation of the deformation mechanism map proposed by *Frost* and *Ashby* [8.4]. N-H indicates the region of Nabarro-Herring creep. $\dot{\varepsilon}_i < \dot{\varepsilon}_{i+1}$

8.1.1 Dislocation Glide

When the applied stress exceeds the ideal strength of about $\mu/20$, any material will be deformed with an unlimited strain rate. This type of deformation occurs in region I in the map. The high-stress region below the ideal strength is region II, dislocation glide. In this region, the deformation rate is sufficiently high that the effect of diffusion on deformation may be disregarded. In this sense, region II is essentially the same as the low-temperature region, and will not be treated in this chapter.

8.1.2 Diffusional Creep

In the region where the stress is sufficiently low for dislocations not to be moved, deformation occurs exclusively by directional diffusion to reduce the applied stress. For this type of deformation, the transient stage is almost absent, i.e., the deformation proceeds in a steady-state creep of constant stress–constant strain rate conditions. This region is called a diffusional creep region.

Why the plastic deformation occurs by self-diffusion is illustrated in Fig. 8.2. Under a uniaxial tensile stress, a normal stress as high as the applied stress σ works on a boundary almost perpendicular to the applied stress, say boundary AB in the figure. When a vacancy is formed there, the formation energy is reduced by $\sigma\Omega$, where Ω is the volume of the vacancy. Accordingly, the concentration of vacancies in thermal equilibrium is given by

$$c_v = c_{v0} \exp\left(-\frac{\Delta H_f - \sigma\Omega}{k_B T}\right). \tag{8.1}$$

Here, c_{v0} is a constant determined by the change in thermal entropy associated with the vacancy formation, ΔH_f the formation enthalpy of the vacancy, k_B the Boltzmann constant, and T the absolute temperature.

Under low-stress and high-temperature conditions where $\sigma\Omega \ll k_B T$, (8.1) is approximated by

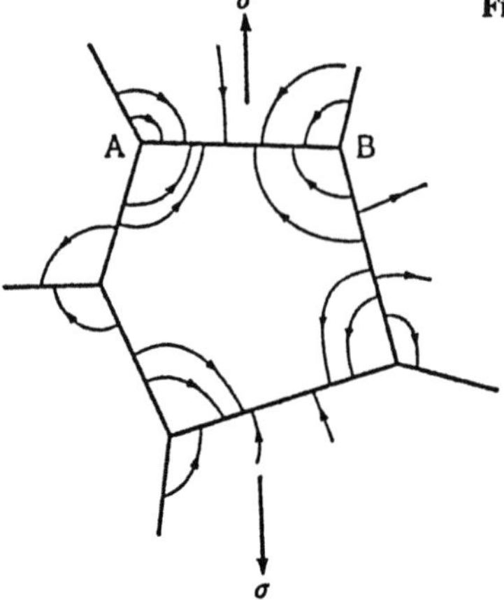

Fig. 8.2. Directional diffusion of atoms under a tensile stress

$$c_v = c_{v0} \exp\left(-\frac{\Delta H_f}{k_B T}\right)\left(1 + \frac{\sigma\Omega}{k_B T}\right).$$

Here,

$$c_{ve} = c_{v0} \exp\left(-\frac{\Delta H_f}{k_B T}\right)$$

is the concentration in thermal equilibrium under no applied stress. It is noted that c_v is larger than c_{ve} by

$$\Delta c_v = c_{ve}\frac{\sigma\Omega}{k_B T}. \tag{8.2}$$

On the boundary almost parallel to the applied stress, the normal stress component σ_n is very low and thus Δc_v is very small. Where $\sigma_n = 0$, $\Delta c_v = 0$. As a result, a concentration gradient is induced between differently inclined boundaries, and vacancies flow directionally. Material flow occurs in the opposite direction to the vacancy flow, as shown in Fig. 8.2, and the crystal elongates in the stress direction.

Near the corners such as A and B, the vacancy flow is larger because of the higher concentration gradient. However, if these regions elongate more than the central part of AB, an elastic internal stress is produced by the inhomogeneous deformation: the internal stress is compressive in places of larger elongation and the stress is tensile in places of smaller elongation. As a result, the elongation is suppressed at the corners and enhanced at the central parts. Therefore, an inhomogeneous elongation larger than σ/E will not be produced. Here, E is Young's modulus. Since σ/E is very small, the elongation becomes practically homogeneous.

Denoting the grain diameter by d, the concentration gradient is approximately equal to $\Delta c_v/d$, because the length of the diffusion path is about d. The flow flux is thus given by

$$J \approx D_v c_{ve} \frac{\sigma \Omega}{k_B T} \frac{1}{d} . \tag{8.3}$$

Here D_v is the diffusion coefficient of vacancies. Supposing that J is expressed by the number of vacancies flowing across unit area per unit time, the dimensions of D_v are [number/(length $\times$ time)] because c_{ve} is dimensionless. Therefore, the volume of vacancies passing through unit area per unit time is $J\Omega$ and the coefficient of lattice self-diffusion D_l is given by $D_v c_{ve}\Omega$. Thus, the elongating rate of the grain is approximately equal to $D_l \sigma \Omega / k_B T d$, and the tensile strain rate is given by

$$\dot{\varepsilon} = A \frac{D_l \Omega}{k_B T} \frac{\sigma}{d^2} , \tag{8.4}$$

where A is a constant that depends on the shape of the grain.

The grain boundaries nearly parallel to the stress axis absorb vacancies and the grains having these boundaries become thinner in the direction perpendicular to the stress axis. As a result, some grain boundary sliding must occur so that cavities are not produced between grains. This grain boundary sliding also contributes to the grain elongation. This component of strain is also dependent on the shape of grains [8.5]. Considering these effects, the value of A is expected to be 12–40, depending on the grain shape [8.6]. The experimental value obtained by tensile test is about 40 in many cases [8.7]. Since this deformation mechanism was proposed by *Nabarro* [8.8] and developed by *Herring* [8.9], it is called *Nabarro-Herring creep* and dominates in region V in Fig. 8.1.

Self-diffusion also occurs along grain boundaries. Since the activation energy for grain boundary diffusion is no more than 0.5–0.6 times that for lattice diffusion, at lower temperatures the grain boundary diffusion dominates over the lattice diffusion. However, the diffusion path is limited to grain boundary regions, and the cross section of the path is smaller than that for lattice diffusion by a factor of w/d (where w is the thickness of the grain boundary). Therefore, the strain rate due to the grain boundary diffusion is given by

$$\dot{\varepsilon} = A \frac{D_{gb} \Omega}{k_B T} \frac{\sigma}{d^2} \frac{w}{d} , \tag{8.5}$$

where D_{gb} is the coefficient of grain boundary diffusion and w is as small as $2b$ in metals, b being the interatomic distance. This deformation mechanism was proposed by *Coble* [8.10], who showed that the value of A is about 50. The deformation controlled by this mechanism is called *Coble creep*, and is dominant in region VI in Fig. 8.1.

Both Nabarro-Herring creep, which is dominant at high temperatures, and Coble creep, which is dominant at lower temperatures, show a linear stress dependence of strain rate. This same stress-dependence means that the boundary between regions V and VI is independent of stress and perpendicular to the abscissa (temperature coordinate) in Fig. 8.1.

8.1.3 Power Law Creep

Between the high-stress/dislocation glide region and the low-stress/diffusional
creep region, there exists a region where both dislocation glide and diffusion
contribute to the deformation. This does not mean that their contributions are
additive, but that although the main carriers of deformation are dislocations,
their motion is controlled by diffusion.

When the dislocation density is increased by deformation, the internal stress
increases and the movement of dislocations is obstructed by their mutual interaction. As a result, the deformation will stop in time under a constant applied
stress unless diffusion occurs. On the other hand, when diffusion occurs, so that
dislocations can climb and the internal stress decreases through rearrangement or
pair annihilation of dislocations, deformation proceeds until the work hardening
compensates the decrease in internal stress. In other words, the glide motion of
dislocations is not a diffusion-controlled process, but the decrease of obstacles to
the glide motion is diffusion controlled. This concept holds when the major obstacle is the internal stress arising from interdislocation interaction (Sect. 8.2.1).
When there exist obstacles such as jogs in screw dislocations and solute atmospheres around dislocations, the glide motion of dislocations is also controlled
by diffusion. In any case, when the recovery of the dislocation structure due to
annealing balances the change in the structure due to deformation, a steady-state
deformation is realized.

The motion of jogs in screw dislocations is climb when they move along with
the screw dislocations, and this region is often called the dislocation-climb region.
This region is also referred to as the power law creep region. This nomenclature
comes from the fact that the steady-state creep rate in this region is proportional
to the nth power of the applied stress as [8.11]

$$\dot{\varepsilon} = AD\frac{\mu b}{k_{\mathrm{B}}T}\left(\frac{\sigma}{\mu}\right)^n .$$ (8.6)

Here D is the coefficient of lattice self-diffusion, D_{l}, in the sufficiently high
temperature region, while it is the coefficient of pipe diffusion along dislocation
cores, D_{p}, in the relatively low temperature region; this difference has the same
cause as described for diffusional creep. The value of D_{p} is approximately the
same as that of D_{gb}.

Denoting the cross section of the dislocation core by S_{c} and the dislocation
density by ϱ, the cross section of the diffusion path is $S_{\mathrm{c}}\varrho$ per unit area. Here, S_{c}
is about b^2. Thus, the contribution of pipe diffusion to dislocation climb is lower
than that of lattice diffusion by the areal ratio $S_{\mathrm{c}}\varrho$. On the other hand, when the
flow stress is solely determined by the interdislocation interaction, the relation

$$\sigma \propto \mu b\sqrt{\varrho}$$ (8.7)

holds, which leads to $S_{\mathrm{c}}\varrho \propto (\sigma/\mu)^2$. Therefore, as long as the deformation
mechanism other than diffusion is the same, the stress exponent n in the lower

temperature (LT) region of pipe diffusion control is larger by 2 than the stress exponent in the higher temperature (HT) region of lattice diffusion control. As a result, the boundary between the lower temperature region III and the higher temperature region IV becomes higher on the right and the lower temperature region expands to the higher temperature side as the stress is increased.

The idea described above cannot be applied to highly solution-hardened alloys. In such alloys, solute atmospheres are formed around dislocations and hinder the dislocation motion. Since this resistance is added, the flow stress becomes larger than that given by (8.7). When the diffusion rate is sufficiently high, this resistance increases in proportion to the dislocation velocity. If the strain rate is the same, the average dislocation velocity decreases in inverse proportion to the dislocation density and this resistance decreases in parallel. Accordingly, it is sometimes observed that σ becomes smaller for higher ϱ. Besides, in such alloys the temperature dependence of the strength behaves in a complicated manner depending on the relative magnitude of the diffusion velocity to the dislocation velocity. For these reasons, the above argument concerning the stress exponent does not necessarily hold.

Equation (8.6) always holds for the alloys mentioned above, though the rate-determining process for high-temperature deformation is now not the climb but the glide motion of dislocations. Consequently, this region should be called, in a more general sense, the power law creep region rather than the dislocation climb region. In the HT region, the stress exponent n is 4–5 in pure metals, about 3 in highly solution-hardened alloys and sometimes becomes as large as 40 in dispersion-hardened alloys.

In the cases where the obstacles to dislocation motion diffuse much faster than solvent atoms, e.g., interstitial solute atoms in bcc metals, the diffusion of solute atoms may affect the deformation even at lower temperatures, where the climb motion of dislocations hardly occurs. These effects are not considered in constructing Fig. 8.1.

Since the diffusion-controlled dislocation motion described in this section is the dominant deformation mechanism under practical deformation conditions, it has been investigated in many studies, which will be described in detail in the following sections.

Equi-strain-rate lines are shown in Fig. 8.1. The region for $\dot{\varepsilon} \lesssim 10^{-10}\mathrm{s}^{-1}$ would more aptly be called an elastic region, because practically no plastic deformation occurs in this region. The Coble creep region appears very wide, but most of it is practically the elastic region.

8.1.4 Harper-Dorn Creep

Harper and *Dorn* [8.12] found a mechanism that is different from either diffusional creep or the usual power law creep. (This mechanism is not included in *Ashby*'s deformation mechanism map.) The constitutive equation is

$$\dot{\varepsilon} = AD_1 \frac{\mu b}{k_\mathrm{B} T} \frac{\sigma}{\mu} , \qquad (8.8)$$

where A is 1.67×10^{11} for aluminum. This kind of deformation resembles Nabarro-Herring (N-H) creep in that the strain rate is proportional to the lattice diffusion coefficient and applied stress, but it is clearly different from N-H creep in that the strain rate is much higher than the theoretical rate of N-H creep, no less than say 1000 times for aluminum, and that no selective elongation is observed at the grain boundaries perpendicular to the tensile axis. Further, the strain rate does not show the grain size dependence inherent in N-H creep. Although it is reported that dislocations have some connection with this deformation [8.12, 13], the mechanism has not yet been clarified.

8.1.5 Effect of Internal Structure

In crystalline materials, the dislocation density is increased by plastic deformation, and work hardening occurs in many cases. Precipitation also yields hardening. When such a change in internal structure occurs, the boundary lines in the deformation mechanism map will shift. For example, the boundary line between Coble creep and dislocation creep in Fig. 8.1 will shift to the higher stress side as the density of obstacles to dislocation motion increases. It is inconvenient that such a shift of boundary line takes splace during deformation. In this respect, the deformation mechanism map is more useful for high-temperature deformation, because a steady or quasi-steady state is realized for the internal structure and the boundary line is uniquely determined for the steady-state structure. Especially for creep deformation, the main part of the strain required for rupture is usually attained by the deformation in the steady state.

For many practical applications, diffusional creep prevails, which is strongly dependent on grain size. In this respect, *Ashby*'s map is inconvenient, because it does not include the effect of grain size. Then *Langdon* and *Mohamed* [8.7, 14] proposed a map which emphasizes the effect of grain size. They also showed that the map is much simplified if the inverse homologous temperature is used instead of homologous temperature.

For every deformation mechanism operating at high temperatures, the strain rate is proportional to powers of stress and grain size, and linearly dependent on diffusion coefficient given by $D = D_0 \exp(-Q/RT)$. Therefore, if the axes for stress and grain size are expressed by their logarithms and the axis for temperature by inverse absolute temperature, linear relationships hold among the three variables for a given strain rate, i.e., the boundaries in a high-temperature deformation mechanism map are expressed as planes in a three-dimensional space. Figure 8.3 is the three-dimensional map given by *Oikawa* [8.15] for pure aluminum. Figures 8.4 and 8.5 are constant-T and constant-σ cross sections given by *Mohamed* and *Langdon* [8.7, 14].

In general, when several independent deformation mechanisms operate simultaneously, the mechanism which gives the highest strain rate is regarded as the controlling mechanism. Then the strain rates for two independent mechanisms become the same on the boundary between the two regions where one of the two mechanisms is the controlling one. The boundary is given by

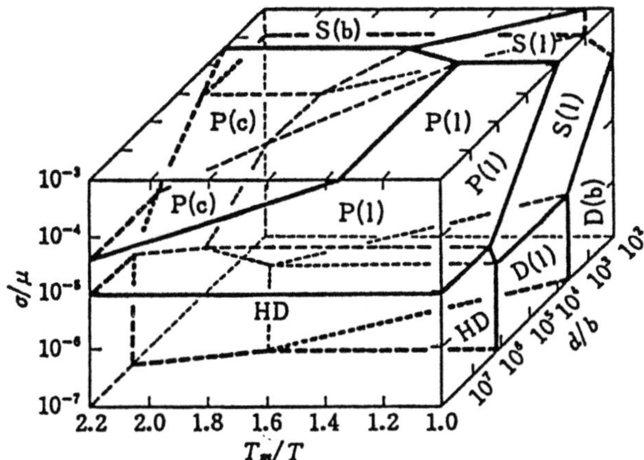

Fig. 8.3. Three-dimensional deformation mechanism map proposed by *Oikawa* [8.15] for pure aluminum.

HD: Harper-Dorn creep; D(l): Diffusional creep (controlled by lattice diffusion); D(b): Diffusional creep (controlled by grain-boundary diffusion); P(l): Power-law creep (controlled by lattice diffusion); P(c): Power-law creep (controlled by dislocation-core diffusion); S(l): Superplasticity (controlled by lattice diffusion); S(b): Superplasticity (controlled by grain-boundary diffusion)

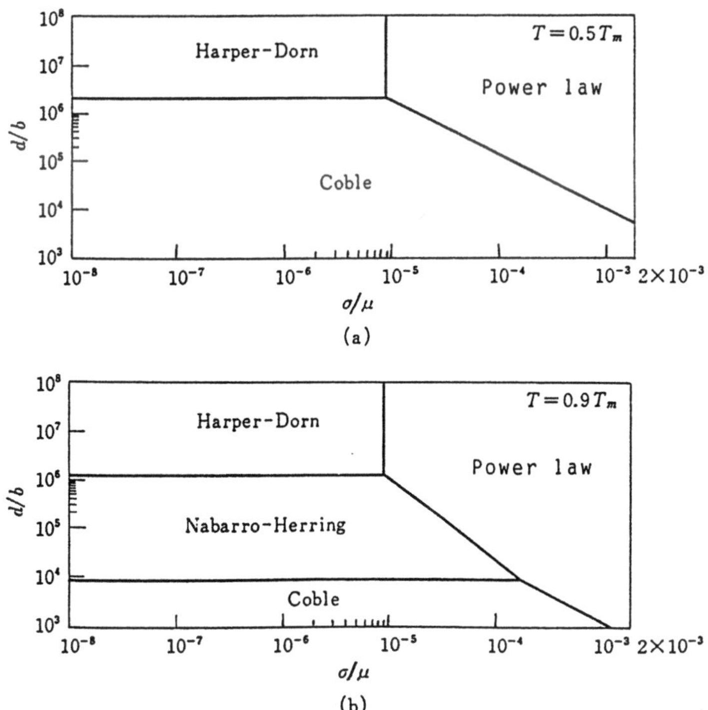

Fig. 8.4a,b. Deformation mechanism map proposed by *Mohamed* and *Langdon* [8.7] for pure aluminum. Cross sections of constant temperatures, (a) $T = 0.5T_m$ and (b) $T = 0.9T_m$

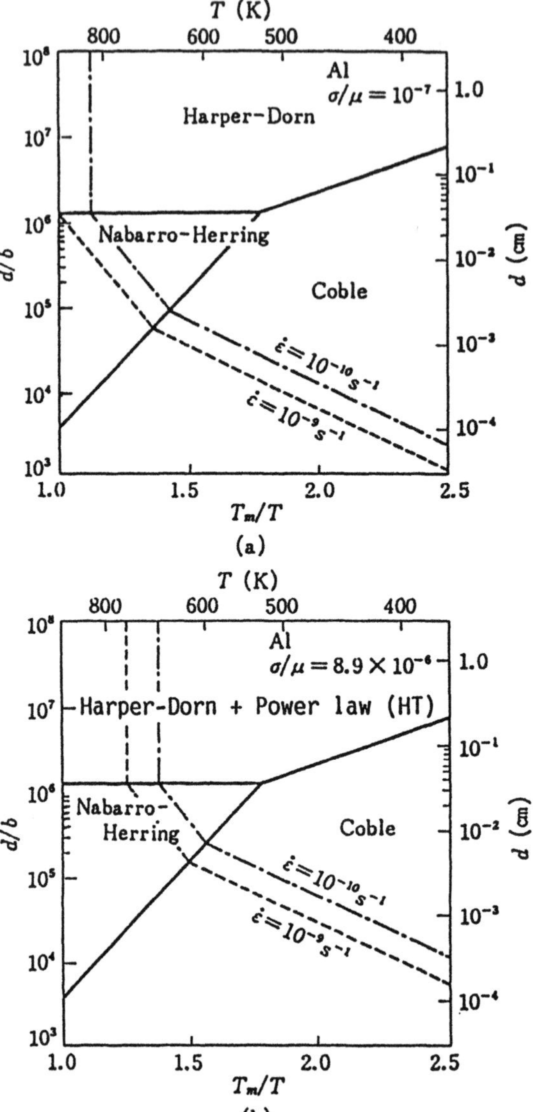

Fig. 8.5a,b. Deformation mechanism map proposed by *Langdon* and *Mohamed* [8.14] for pure aluminum. Cross sections of constant stresses, (a) $\sigma = 10^{-7}\mu$ and (b) $\sigma = 8.9 \times 10^{-6}\mu$

$$\Delta n \log\left(\frac{\sigma}{\mu}\right) + \Delta p \log\left(\frac{d}{b}\right) - \frac{\Delta Q}{2.3 R T_m}\frac{T_m}{T} = \text{const} , \qquad (8.9)$$

where Δn is the difference in stress exponent between the two mechanisms, Δp the difference in grain size exponent, ΔQ the difference in activation energy, and R the gas constant.

For example, power-law creep of higher temperature type and Coble creep are thought to be independent of each other. By equating (8.5) and (8.6), we obtain $\Delta n = n - 1$, $\Delta p = 0 - (-3) = 3$, $\Delta Q = Q_1 - Q_{gb}$, and const = $\log(A_C D_{0(gb)} w \Omega / A_{HT} D_{0(l)} b^4)$, where the suffix C means Coble creep, HT power-law creep of higher temperature type, l lattice diffusion, and gb grain boundary diffusion.

The parameter $\log(\sigma/\mu)$ is used as the stress axis, which is the same scheme as *Ashby* used. The grain diameter is normalized by the interatomic distance to be dimensionless. As mentioned before, Q_{gb} is about $(0.5-0.6)Q_1$, and Q_1 is linearly dependent on the melting point T_m. Therefore, the inverse of homologous temperature, T_m/T, is used as the temperature axis for generalization. In this representation, the slope of the boundary is given by $\Delta Q/2.3RT_m \Delta p$ on a constant-σ cross section, and it is given by $-\Delta n/\Delta p$ on a constant-T cross section. Accordingly, once a point is found on the boundary, the boundary line will easily be determined. Further, the equi-strain-rate lines are also given as straight lines, as is evident from the constitutive equations. When the stress exponent is the same but the grain size exponent differs across the boundary, e.g., between Harper-Dorn, Nabarro-Herring and Coble creeps, the boundary separating them is parallel to the stress axis. Conversely, when the grain size exponent is the same and the stress exponent differs across the boundary, e.g., between Harper-Dorn and power-law creeps, the boundary is perpendicular to the stress axis.

8.1.6 Others

a) The Region of Superplasticity. Fine-grained materials are often observed to be elongated over several hundred per cent by tensile deformation, which is called superplasticity. Superplasticity is thought to be realized by grain-boundary sliding. *Lüthy* et al. [8.16] proposed to incorporate this mechanism in the map, assuming that the mechanism is independent of either diffusional creep or power-law creep. Summarizing the previous studies, they showed that the constitutive equation is described by

$$\dot{\varepsilon} = AD_{gb}\frac{\mu b}{k_B T} \left(\frac{b}{d}\right)^3 \left(\frac{\sigma}{\mu}\right)^2 \tag{8.10}$$

for grain-boundary diffusion control, and

$$\dot{\varepsilon} = AD_l\frac{\mu b}{k_B T} \left(\frac{b}{d}\right)^2 \left(\frac{\sigma}{\mu}\right)^2 \tag{8.11}$$

for lattice diffusion control, and gave a map as shown in Fig. 8.6. Here, A is about 2×10^5 for grain-boundary diffusion and about 8×10^6 for lattice diffusion. The region of superplasticity proposed by *Lüthy* et al. also appears in the three-dimensional map given by *Oikawa* (Fig. 8.3).

The temperature and grain size dependence of the strain rate for superplasticity are the same as those for Coble creep in the region of grain-boundary diffusion and those for N-H creep in the region of lattice diffusion. However, the stress exponent is larger than that for diffusional creep. Therefore, the mechanism

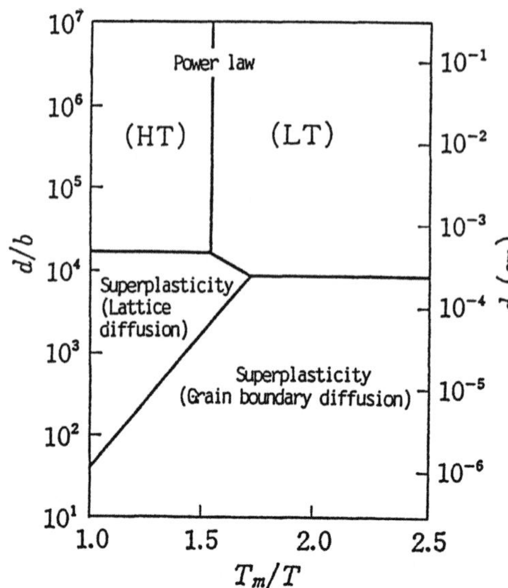

Fig. 8.6. Deformation mechanism map incorporating the superplasticity region proposed by *Lüthy* et al. [8.16] for pure aluminum. Cross section of constant stress, $\sigma/\mu = 5 \times 10^{-4}$

prevails in a higher stress region than diffusional creep, and is situated between power-law creep and diffusional creep.

b) Independence of Deformation Mechanism. The grain-boundary sliding mentioned above is observed not only in the superlasticity region but also in the power-law creep and diffusional creep regions. In the power-law creep region, however, the relative contribution, $\varepsilon_{gb}/\varepsilon_t$, of the strain due to grain boundary sliding ε_{gb} to the total strain ε_t is less than 0.5 and independent of strain, though it depends on stress [8.13]. The strain independence of the ratio suggests that ε_{gb} is determined by the strain in grains, and thus the deformation in grains and the grain-boundary sliding are not independent deformation mechanisms. Some of the dislocations that have contributed to the grain interior deformation enter grain boundaries and climb-glide along the boundaries. In this case, the grain-boundary sliding is thought to occur by the shear component of the deformation arising from the climb-glide motion.

The power-law creep is considered to be controlled by the climb motion of dislocations. As the creep deformation occurs so that the work hardening compensates the recovery-softening caused by the climb motion, the glide and climb motions of dislocations are not independent of each other. In the power-law creep region of highly solution-hardened alloys, the case sometimes occurs where the glide velocity of a dislocation dragging a solute atmosphere is reduced to the climb velocity, so that the glide motion controls the deformation [8.17].

As seen in the above examples, where the deformation mechanisms are not independent of each other but are sequential processes, the boundary separating the regions can no longer be determined by the constitutive equations of the two mechanisms.

c) Equi-strain-rate Curve. The equi-strain-rate curves shown before in the deformation mechanism map bend sharply at the boundaries, but this is an approximate expression. When two independent deformation mechanisms operate simultaneously, the strain rate is doubled at the boundary, because the rate is given by the sum of the strain rates due to the two mechanisms. Accordingly, a true equi-strain-rate curve smoothly bends at a boundary [8.7]. However, as the curved region is narrow, this effect is, in practice, neglected.

8.2 Deformation due to Dislocation Motion

When the deformation is carried by dislocations, the flow stress is determined by the resistance to dislocation motion. The cause of work hardening or softening is that the density and distribution of the obstacles to dislocation motion or the mobile dislocation density undergoes a change due to deformation. For low-temperature deformation, we should consider only the change due to deformation, such as the increase in dislocation density and point defects or the refinement of precipitates. For high-temperature deformation, however, we should consider, in addition, that the annealing effects, such as recovery and change in precipitates, take part in the deformation. These effects, particular to high temperature, make it difficult to identify the obstacle to deformation. In this section, the relation between the type of obstacles and the characteristics of deformation resistance will be described, and why the identification becomes difficult at high temperature and the method to overcome the difficulty will be discussed.

8.2.1 Thermal and Athermal Processes

Various kinds of resistances to dislocation motion can be considered. Consider first a fully annealed crystal, where dislocations form a network structure. When link dislocations lying on a slip plane bow out at the onset of deformation, they are resisted by their own line tension. Even after they free themselves from this initial restraint, other resistances work on them: Peierls barrier; the passing of oppositely signed dislocations; cutting with forest dislocations; and dragging of jogs in screw dislocations. In alloys the resistance due to solute atoms or precipitates or both is added.

There are two processes, thermal and athermal, for a dislocation to overcome a barrier. Which of the two processes is realized depends on the kind of barrier. In the case of jog-drag motion of a screw dislocation, for example, which is shown in Fig. 8.7, the dislocation should form or absorb a vacancy at a jog for every motion by b. Further, the formed vacancies should diffuse away. Thus resistance ranges as short as 1 atomic distance and at higher temperatures can be overcome by a thermally activated process.

Denoting the obstacle spacing by l and the resolved shear stress by τ_e, the work done by the stress is given by $l\tau_e b d_a$ for an activation process, where d_a is

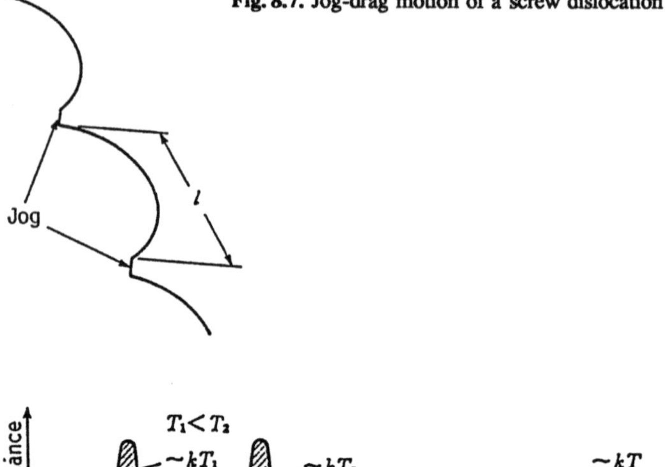

Fig. 8.8. (a,b) Temperature-dependent resistance and (c) almost temperature-independent resistance [8.18]

the range of resistance shown in Fig. 8.8. The parameter $lbd_a = V_a$ is referred to as the activation volume and $ld_a = A_a$ as the activation area.

When the mobile dislocation density is ϱ_m, the number of activation points per unit volume is given by ϱ_m/l. Denoting the frequency factor by ν (including an entropy term) and the area swept by a dislocation per event of activation process by S, the shear strain rate is given by

$$\dot{\gamma} = \frac{\varrho_m}{l}\,\nu bS \exp\left(-\frac{U_0 - V_a\tau_e}{k_B T}\right) , \tag{8.12}$$

where U_0 is the activation energy for $\tau_e = 0$. This equation can be rewritten as

$$\tau_e = \frac{U_0 - k_B T \ln(\dot{\gamma}_0/\dot{\gamma})}{V_a} , \tag{8.13}$$

where

$$\dot{\gamma}_0 = \frac{\varrho_m \nu bS}{l} .$$

For a given strain rate, as the temperature rises, the energy of thermal vibration increases and the work to be done by the stress can be reduced by that amount and the resistance decreases. In other words, when the term $(U_0 - V_a\tau_e)/k_B T$ in (8.12) is a constant, τ_e decreases with the increase in T. Further, for an increasing strain rate, it is necessary to decrease the activation energy, by increasing the stress.

The above mentioned temperature and strain-rate dependence of flow stress is considered always to occur when a dislocation can surmount an obstacle by a thermally activated process. On the other hand, when the resistance is of long range, as for internal stress arising from other dislocations, the flow resistance is hardly affected by the thermal vibrations, as illustrated in Fig. 8.8c. For a dislocation to overcome an obstacle with an observable frequency, the dislocation should be pushed up very near to the top of the peak of the barrier. Only a slight decrease in stress drastically increases the activation energy (shown as a hatched area in the figure). Accordingly, the stress necessary to make the dislocation surmount the obstacle is almost unchanged by varying temperature and strain rate. This is the case for very large U_0 and V_a, and the deformation is considered to be athermal. Even in the case of a short-range obstacle, if

$$U_0 \gg k_B T \ln \frac{\dot{\gamma}_0}{\dot{\gamma}} \,,$$

the value of τ_e is almost determined by U_0/V_a; although τ_e varies with temperature and strain rate because V_a is small, the variation becomes negligibly small compared with U_0/V_a. Therefore, if U_0 is sufficiently large compared with $k_B T$, the process can be regarded as athermal.

The athermal resistance, τ_i, due to interdislocation interaction is expressed as

$$\tau_i = \alpha \mu b \sqrt{\varrho} \,, \tag{8.14}$$

where ϱ is the dislocation density and α is a constant depending on dislocation arrangement. If the distribution of dislocations is uniform, α is about 1/15 for dislocations passing each other, about 1/3 for interaction with forest dislocations and about 1 for bowing out of a dislocation loop as in the case of a Frank-Read source.

When a thermal stress τ_e coexists with the athermal stress, the flow stress is given by the sum of these stresses:

$$\tau = \tau_e + \tau_i \,. \tag{8.15}$$

Here, $\tau_e = \tau - \tau_i$ is sometimes called the effective stress (Sect. 2.2.1), because it is the stress which effectively works on the thermally activated motion of dislocations. τ_i is sometimes called internal stress from the point of view that it arises from the internal stress of other dislocations. The resistance to the bowing out of a dislocation loop, however, comes from its own line tension and not an internal stress. In contrast to τ_e, τ_i may be called athermal stress. Such a name, however, seems to give the wrong impression that it is the stress itself that can in some cases be activated and in others not. Therefore, τ_i is called internal stress in this text.

8.2.2 Viscous Motion and High-Speed Motion of Dislocations

In Sect. 8.2.1, the case where obstacles can be regarded as stationary ones that resist the movement of dislocations was described. Now consider the case where the obstacles can move, interacting with dislocations. As will be described in detail in Sect. 9.2, when the temperature is sufficiently high to allow the diffusion of solute atoms, a solute atmosphere is formed even around moving dislocations and dislocations move viscously dragging the atmosphere. The dragging resistance depends on the magnitude of the velocity of solute atoms induced by the interaction with the dislocation relative to the dislocation velocity. At high temperatures and low strain rates, where the relative velocity is sufficiently high, the resistance decreases with rising temperature or with decreasing strain rate. In other words, the solute atmosphere dragging can be described as a thermally activated process as is the case for stationary obstacles.

The jog dragging of screw dislocations described in Sect. 8.2.1 is also viscous-like, because the dislocation must overcome obstacles for every movement of b and the motion over a distance much longer than b looks viscous. In general, when activation points are distributed with an average spacing l on a slip plane, once a dislocation being stopped by an obstacle overcomes the obstacle, it moves by about l. Over a much longer distance than l, the repeated stops and sudden movement can be regarded as quasi-viscous motion with a velocity

$$v \cong l\nu \exp\left(-\frac{U_0 - V_a\tau_e}{k_B T}\right). \tag{8.16}$$

In other words, the dislocation motion is in general viscous if an effective stress exists. In this case, the average velocity $\bar{v}$ of dislocations can be unambiguously defined, and the strain rate is given by the Orowan equation,

$$\dot{\gamma} = \varrho_m b\bar{v}. \tag{8.17}$$

On the other hand, in the case of long-range and athermal resistance, once a dislocation overcomes an obstacle with the aid of applied stress, it moves very fast to the next obstacle. This kind of movement is often called free flight motion. At high temperatures, however, dislocations experience a phonon resistance. Therefore, the motion is not literally free flight. It is called here high-speed motion.

When the high-speed flight distance is long, it should be taken into account that during a flight the dislocation length changes and thus the dislocation density changes. Therefore, contrary to the case of thermally surmountable obstacles where the average dislocation velocity can be defined, the strain rate should be expressed as

$$\dot{\gamma} = \dot{N} b\overline{S}, \tag{8.18}$$

where $\dot{N}$ is the number of dislocation loops overcoming obstacles per unit volume per unit time and $\overline{S}$ is the average area swept out by a loop per event.

If no recovery effect exists, as deformation proceeds, the dislocation density increases and the density of the athermal obstacles on the path of dislocation increases along with the strength of their resistance. Therefore, in order to continue the deformation, the applied stress should be increased. This means that the yielding rate $\dot{N}$ of mobile dislocation loops is determined by the internal structure such as dislocation arrangement and by the increasing rate of applied stress, $\dot{\tau}$. On the other hand, at high temperatures where recovery or recrystallization proceeds rapidly, dislocation rearrangement and annihilation easily occur and the strength as well as the density of athermal obstacles decreases. This decreasing rate also contributes to $\dot{N}$. For creep deformation proceeding under a constant applied stress, the decreasing rate just determines $\dot{N}$.

8.3 Identification of Deformation Mechanism at High Temperatures

The flow stress of metals and alloys depends on temperature in general as shown in Fig. 8.9. In the low-temperature region A and in the high-temperature region C, the flow stress decreases as temperature rises. Therefore, the deformation in these regions should be controlled by some thermally activated process.

In the intermediate temperature region B, the flow stress is almost independent of temperature in some cases, and it increases with the rise of temperature in the other cases. This inverse temperature dependence is observed only in those alloys which can be hardened by strain aging. This suggests that the obstacles to dislocation motion should be formed by a thermally activated process.

Since the high-temperature deformation described in this chapter should be a thermally activated process described above, it seems most probable that some effective stress contributes to the flow stress together with an internal stress. If the effective stress and the internal stress can be determined separately, the activation energy and activation volume for the deformation are determined from the temperature and strain-rate dependences of the effective stress, and the obstacle responsible for the thermal resistance may be inferred. From such a standpoint,

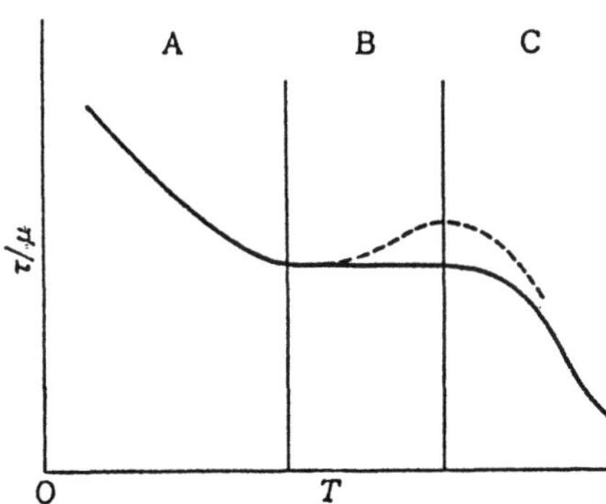

Fig. 8.9. Temperature dependence of modulus compensated flow stress τ/μ

135

several techniques have been proposed to separate the effective and internal stresses.

Many of the techniques are based on the assumption that if the temperature or strain rate is changed under such conditions that the internal structure does not change significantly, the change in effective stress can readily be found from the change in flow stress. It should be noted that this assumption is acceptable only when the internal stress for a given internal structure is independent of temperature and strain rate.

8.3.1 Temperature Change Technique

Since it was proposed by *Seeger* [8.19], this method has been widely used for the separate determination in the low-temperature region (A).

As seen from (8.13), $\tau_e = 0$ in the temperature region

$$T \geqq T_c = \frac{U_0}{k_B \ln(\dot{\gamma}_0/\dot{\gamma})} \ . \tag{8.19}$$

Therefore, once the temperature dependence of flow stress has been determined as shown in Fig. 8.9, the internal stress τ_i in region A can be determined as the flow stress in region B where $\tau_e = 0$, because the internal stress does not depend on temperature.

Since the yield stress in a well-annealed crystal is the flow stress at a very small amount of plastic deformation, the internal structure at the stress is assumed almost independent of temperature. In a work-hardened material, the internal structure would also be almost independent of temperature in the low-temperature region where recovery hardly occurs. This method, however, may not be applied to the high-temperature region where recovery occurs severely.

8.3.2 Strain Rate Change Technique

In this method, the strain rate is changed during deformation and the accompanying change in flow stress is measured, as illustrated in Fig. 8.10.

When a dislocation overcomes stationary obstacles by a thermally activated process, the dislocation velocity may in general be given by (8.16). As the acti-

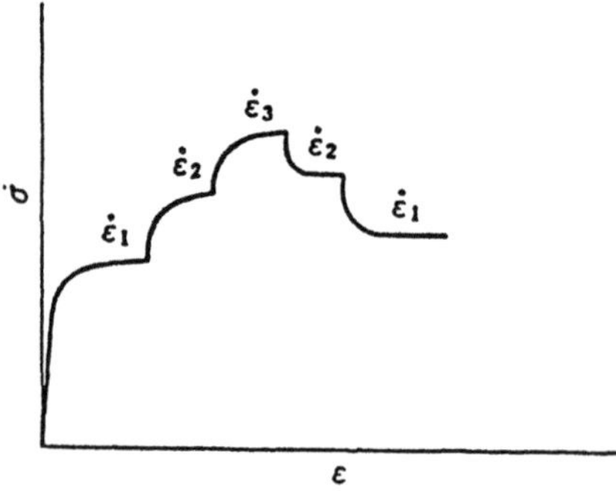

Fig. 8.10. Schematic representation of strain-rate change test conducted with three strain rates, $\dot{\varepsilon}_1 < \dot{\varepsilon}_2 < \dot{\varepsilon}_3$

vation volume V_a is generally a function of stress, the effective stress dependence of dislocation velocity is complicated. However, the velocity is approximately expressed by

$$v = B\tau_e^m = B(\tau - \tau_i)^m \ . \tag{8.20}$$

Li [8.20] showed that the internal stress τ_i and effective stress exponent m can easily be obtained from this approximate equation and the Orowan equation (8.17) by measuring flow stresses τ_1, τ_2, τ_3 corresponding to three strain rates $\dot{\gamma}_1$, $\dot{\gamma}_2$ and $\dot{\gamma}_3$, respectively. If the geometrical average $\sqrt{\dot{\gamma}_1\dot{\gamma}_3}$ of $\dot{\gamma}_1$ and $\dot{\gamma}_3$ is chosen for $\dot{\gamma}_2$ and assuming that ϱ_m and τ_i do not change, the internal stress is obtained as

$$\tau_i = \frac{\tau_1\tau_3 - \tau_2^2}{\tau_1 + \tau_3 - 2\tau_2} \tag{8.21}$$

and the stress exponent as

$$m = \frac{\ln(\dot{\gamma}_2/\dot{\gamma}_1)}{\ln[(\tau_2 - \tau_i)/(\tau_1 - \tau_i)]} = \frac{\ln(\dot{\gamma}_3/\dot{\gamma}_2)}{\ln[(\tau_3 - \tau_i)/(\tau_2 - \tau_i)]} \ . \tag{8.22}$$

The above two methods have a problem in the assumption that the change in internal structure is negligible with the change of temperature or strain rate. Even for the yield stress, the plastic strain is as large as 10^{-3} and the change in dislocation structure due to the strain may not be negligible. In particular, when the local fluctuation of dislocation structure is large, weak regions are at first deformed and work hardened. If very weak regions exist in a very small volume fraction, a large work hardening will occur at an early stage of macroscopic deformation. Even if the deformation of a specimen as a whole is very small, the local internal structure changes remarkably in the weakest region that determines the flow stress. For most annealed materials, the elastic limit does not clearly appear, and the work hardening rate is extremely high at the very beginning of plastic deformation. The high work hardening is thought to come from the above-described local deformation.

As the deformation proceeds, weak regions will be consumed and the internal structure homogenized. Therefore, a technique that is often used is one that suddenly changes temperature or strain rate during deformation, after giving a large deformation. However, in this case also, there is a problem of whether or not the structural variation occurring during the sudden change is sufficiently small. Further, it is not clear which point on the stress–strain curve should be taken as the flow stress for the new deformation condition. If the flow stress at a strain, say 10^{-3}, given under the new condition is taken, the structural change introduced by the strain comes into question, as is the case for yield stress. It would seem likely that the structural change due to a strain as small as 10^{-3} is negligible if it occurs after a large deformation. However, this is not necessarily the case for high-temperature deformation where the recovery effect is very large.

The recovery rate becomes higher with an increase of either temperature or internal stress. On the other hand, the increasing rate of internal stress due to plastic deformation becomes higher with increasing strain rate. Accordingly, the internal stress determined by the competition of the two effects decreases with rising temperature and increases with increasing strain rate. In other words, the internal stress (which is in itself independent of temperature and strain rate, provided that the internal structure is the same) depends on temperature and strain rate for high-temperature deformation through the change in internal structure. Moreover, the manner of the dependence is the same as that of effective stress. This makes discrimination between the two stresses difficult. Therefore, in the case of negligible effective stress it is possible to make the mistake that some effective stress exists.

It usually takes an appreciable time to change temperature, during which some significant change in internal structure is inevitable. Further, with a sudden change of strain rate, the competitive relation between work hardening and recovery also undergoes a sudden change, and it is probable that a significant change in internal structure occurs even during a strain as small as 10^{-3}.

8.3.3 Stress Dip Technique

This method is based on the idea that if $\tau_e > 0$ or $\tau > \tau_i$, the deformation proceeds forward, but if $\tau \leq \tau_i$, the deformation stops or proceeds backward because $\tau_e \leq 0$. Then the applied stress required just to make $\dot{\gamma} = 0$ is equal to the internal stress.

The simplest method of obtaining the applied stress for which $\dot{\gamma} = 0$, is to arrest the cross-head of the test machine during deformation and measure the stress relaxation curve [8.20, 21].

The deformation obtained from the cross-head displacement consists of the elastic deformation of the machine and specimen and the plastic deformation of the specimen. The increment of the cross-head displacement dL is given by the sum of the increments of elastic displacement dl_e and plastic displacement dl_p,

$$dL = dl_e + dl_p .\tag{8.23}$$

If the spring constant of the machine is k_M, the increment of elastic displacement of the machine is given by

$$dl_M = dF/k_M \tag{8.24}$$

for the increment of applied load dF. Denoting the specimen cross section by S_s, the specimen length by l_s, and the Young's modulus by E, the stress increment is expressed as $d\sigma = dF/S_s$, and the increment of elastic elongation of the specimen is

$$dl_{se} = \frac{l_s d\sigma}{E} = \frac{l_s dF}{S_s E} .\tag{8.25}$$

Thus

$$dl_e = dl_M + dl_{se} = l_s d\sigma \left(\frac{S_s}{l_s k_M} + \frac{1}{E} \right) . \tag{8.26}$$

Putting $d\varepsilon_a = dL/l_s$, $d\varepsilon_e = dl_e/l_s$, and $d\varepsilon = dl_p/l_s$, (8.23) is rewritten as

$$d\varepsilon_a = d\varepsilon_e + d\varepsilon ; \tag{8.27a}$$

or dividing both sides of (8.27a) by the time increment dt,

$$\dot{\varepsilon}_a = \dot{\varepsilon}_e + \dot{\varepsilon} . \tag{8.27b}$$

Putting

$$\frac{1}{K} = \frac{S_s}{l_s k_M} + \frac{1}{E} , \tag{8.28}$$

we obtain

$$\dot{\varepsilon}_e = \frac{\dot{\sigma}}{K} . \tag{8.29}$$

$\dot{\varepsilon}_a$ is the apparent strain rate including the elastic deformation of the machine and specimen and usually this strain rate is under control. $\dot{\varepsilon}_e$ is the elastic component of $\dot{\varepsilon}_a$, and K the apparent Young's modulus. $\dot{\varepsilon}$ is the true plastic strain rate and even when the cross-head is arrested so that $\dot{\varepsilon}_a = 0$, $\dot{\varepsilon}$ is not zero, but

$$\dot{\varepsilon} = \frac{-\dot{\sigma}}{K} , \tag{8.30}$$

i.e., the specimen is plastically deformed with a rate proportional to the stress relaxation rate $-\dot{\sigma}$. Therefore, the time when $\dot{\varepsilon} = 0$ is the time when $\dot{\sigma} = 0$ as shown in Fig. 8.11, and the stress level, which does not relax further, is equal to σ_i at that time [8.21, 22].

Usually it takes a long time for the stress to relax to $\dot{\sigma} = 0$. Therefore, a significant change in internal structure due to recovery is inevitable and the internal stress will change. To suppress this change, the stress for $\dot{\sigma} = 0$ should be measured as quickly as possible. For this reason *MacEwen* et al. proposed the incremental unloading test [8.23]. As illustrated in Fig. 8.12, in this technique $\dot{\sigma}$ is examined a short time after an unloading and if $\dot{\sigma} < 0$ the load is further

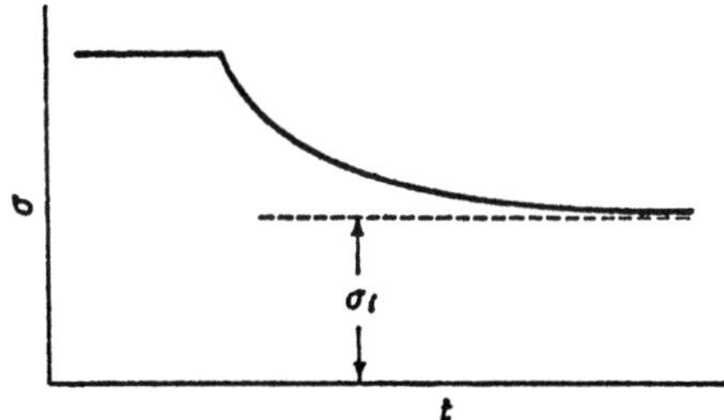

Fig. 8.11. Determination of the internal stress from the condition $\dot{\sigma} = 0$ by the stress relaxation test

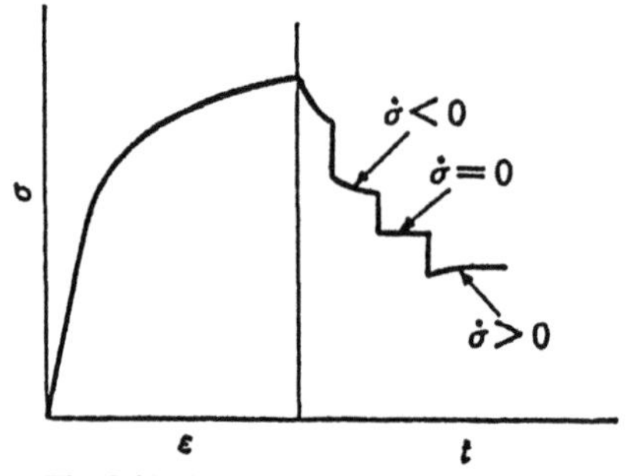
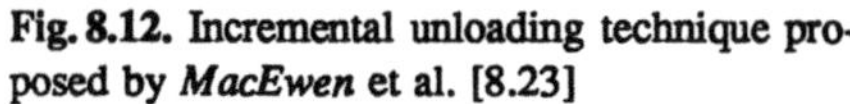

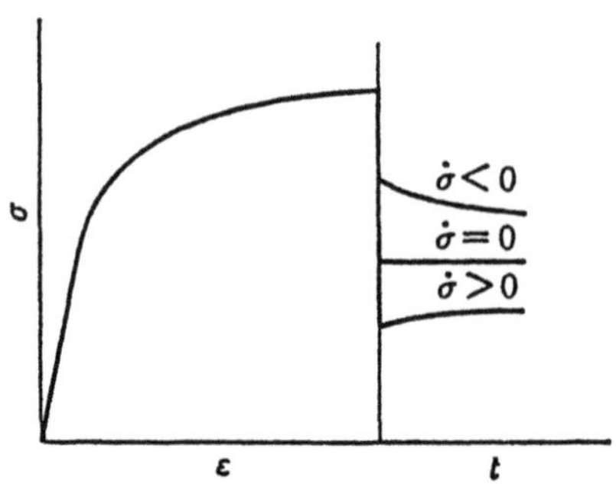

Fig. 8.12. Incremental unloading technique proposed by *MacEwen* et al. [8.23]

Fig. 8.13. Stress dip technique proposed by *Solomon* [8.24]

decreased. This technique allows the required time to be greatly reduced compared with simple stress relaxation. However, some recovery effect may still be included. Accordingly, a technique was proposed in which the stress is rapidly reduced by a large amount at a time.

For specimens in the same deformed state, the applied stress is rapidly reduced to various levels and $\dot{\sigma}$ is examined immediately after the reduction. The stress level for $\dot{\sigma} = 0$, determined in this way, gives the internal stress in the very short time required for only one stress reduction. Figure 8.13 shows a schematic illustration. This technique was proposed by *Solomon* [8.24] and is called the dip test.

When a creep machine is used in place of a tensile machine, $\dot{\varepsilon}$ is directly measured, since $\dot{\sigma} = 0$ in the above has the same meaning as $\dot{\varepsilon} = 0$. This technique, called the strain transient dip test because what is now measured is not the change in σ but the change in ε, was first used by *Ahlquist* and *Nix* [8.25, 26].

When using an electrically controlled hydraulic machine that has recently been developed, the stress can be reduced in a time as short as 2×10^{-2} s. For a conventional machine of the Instron type, however, the screws of the machine must be rotated in the reverse direction to reduce the stress, and a longer time is required. If the speed of reverse rotation is made too high, some inertia effect is often introduced at the time the cross-head stops [8.27], and the rate of stress reduction is limited. The time required for stress reduction depends on the machine and the amount of reduction, and is usually about 1s.

Another problem is how to measure $\dot{\sigma}$ or $\dot{\varepsilon}$ immediately after the stress reduction. The use of a slow response recorder does not allow precise measurements. Further, the assumption that $\sigma = \sigma_i$ when $\dot{\sigma} = 0$ is not correct for high-temperature deformation. When $\sigma = \sigma_i$, $\dot{\sigma} = \dot{\sigma}_i$ and $\dot{\sigma}_i < 0$ in the case of recovery being in progress. In other words, the stress level for $\dot{\sigma} = 0$ is not equal to σ_i.

Figure 8.14 is an example obtained by *Ahlquist* and *Nix* [8.26], where it seems that it takes several seconds or more to judge the sign of $\dot{\varepsilon}$ immediately after the stress reduction. On the other hand, *Toma* et al. [8.27] reported that the stress change measured with a fast response recorder is such as in Fig. 8.15. Figure 8.16

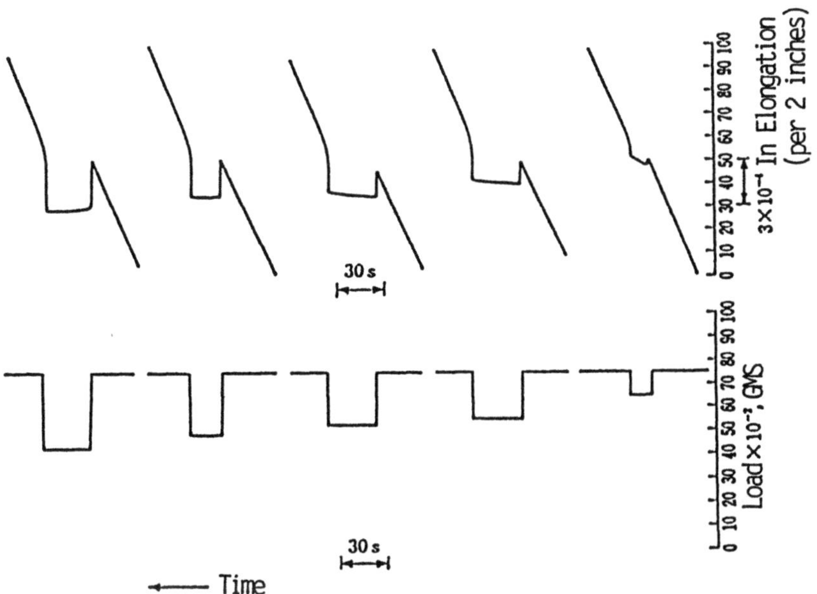

Fig. 8.14. Strain transient after stress dip during steady-state creep at 573 K under 6.8 MPa [8.26], pure aluminum

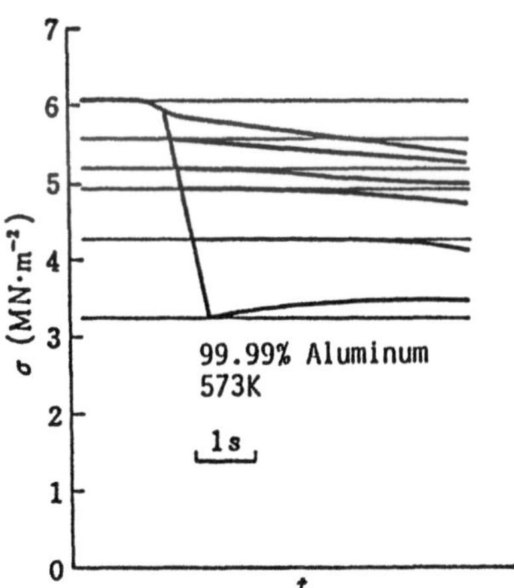

Fig. 8.15. An example of the stress dip test [8.27]

shows the stress–time curves simultaneously recorded with a recorder of response speed of 30 Hz, and one with a transient memory having digital intervals of 10 μs on the time axis and 0.5% of full scale on the stress axis. As seen from these figures, even when $\dot{\sigma} = 0$, immediately after the stress reduction, $\dot{\sigma}$ becomes negative after several seconds. Further, the time interval maintaining $\dot{\sigma} = 0$, Δt, becomes shorter when the stress immediately after the reduction, σ', is higher. Accordingly, *Toma* et al. measured Δt as a function of σ'/σ and the result is

141

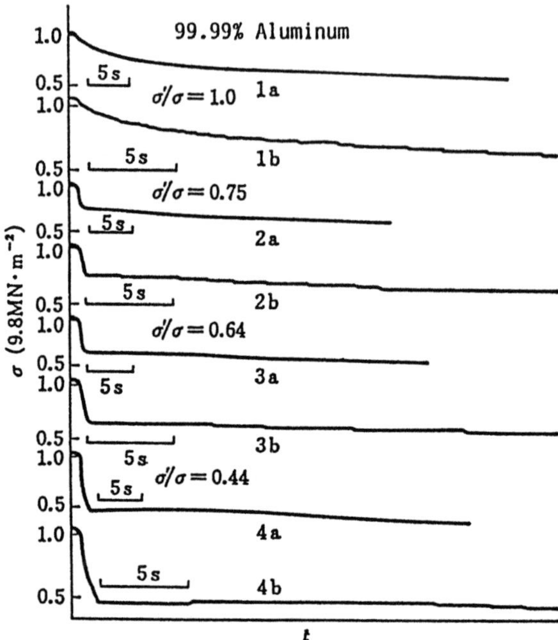

Fig. 8.16. Stress transient after stress dip during steady-state tensile deformation of pure aluminum at 573 K and $3.3 \times 10^{-4}\,\mathrm{s}^{-1}$ [8.27]. Curves denoted by a were recorded by a Linear Corder and curves denoted by b, by a transient memory instrument

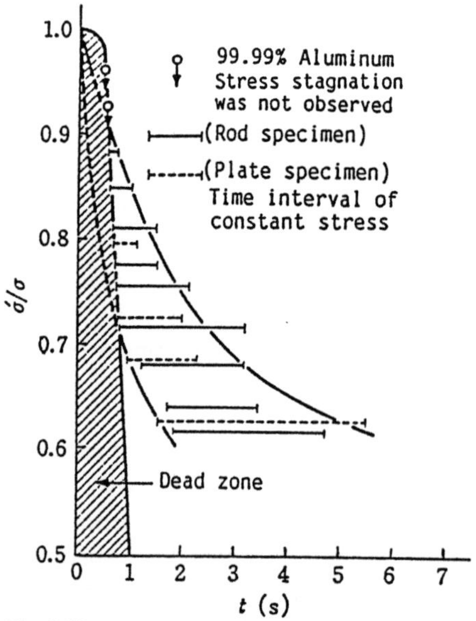

Fig. 8.17. An example of an extrapolation technique. Stress was dipped during steady-state tensile deformation of pure aluminum at 573 K and $3.3 \times 10^{-5}\,\mathrm{s}^{-1}$ [8.27]. $\sigma = 6.1\,\mathrm{MN\,m}^{-2}$

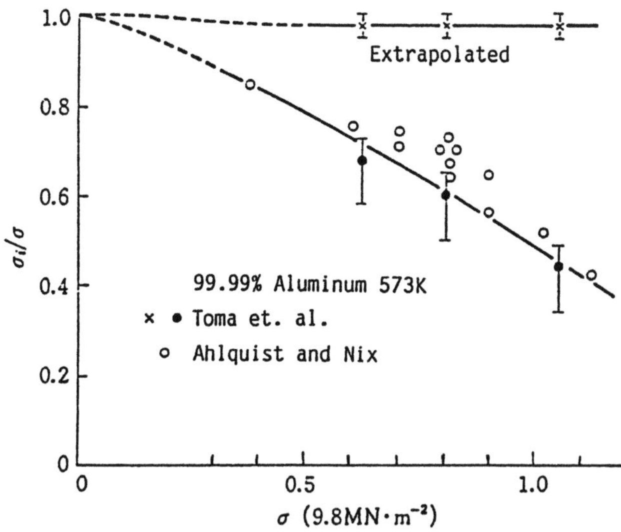

Fig. 8.18. Comparison of the results obtained by *Ahlquist* and *Nix* [8.26] and *Toma* et al. [8.27]

shown in Fig. 8.17. The hatched area in the figure indicates the time needed to reduce the stress. The data points with a downward-pointing arrow show that $\dot{\sigma} < 0$ at that σ' immediately after the stress reduction. The time interval for $\dot{\sigma} = 0$ is considered to be the time for $\sigma' < \sigma_i$, that is, the time elapsed for σ_i to decrease to σ' due to recovery. Thus the value of σ' equal to σ_i immediately before the stress reduction is inferred to be the value of σ' at the end time for $\dot{\sigma} = 0$ extrapolated to $\Delta t \to 0$.

A comparison of the results obtained by *Ahlquist* and *Nix* with those obtained by *Toma* et al. is shown in Fig. 8.18. While the results of *Toma* et al. obtained by judging $\dot{\sigma}$ at several seconds or more after the reduction agree well with those of *Ahlquist* and *Nix*, the internal stress obtained by the extrapolation is almost equal to σ immediately before the reduction and differs significantly from the result of *Ahlquist* and *Nix*. This suggests that the extrapolation is necessary in order to obtain the internal stress by dip tests.

The extrapolation method described above also has a problem concerning the limitation of experimental sensitivity, because the experimental end time for $\dot{\sigma} = 0$ or $\dot{\varepsilon} = 0$ depends on the experimental sensitivity of ε. Since $d\varepsilon = -d\sigma/K$, the resolution of ε can be increased by raising K. The resolution of ε was about 10^{-5} in the case of Toma et al.

When the effective stress is negligible, and in practice $\sigma = \sigma_i$, the fact that $\sigma = \sigma_i$ can be proved more convincingly by another technique, described in Sect. 8.3.6. Further, in the case of $m = 1$ (8.20), σ_i can be measured more precisely by another stress change technique described in the next section. The experimental results obtained by these techniques agree well with the results obtained by the above-mentioned extrapolation method, which shows that the perceived problem of experimental sensitivity does not exist.

8.3.4 Stress Change Technique

When it is assumed that the effective stress dependence of the dislocation velocity is given by (8.20), the tensile strain rate is expressed by

$$\dot{\varepsilon} = \phi^{m+1} \varrho_{\mathrm{m}} b B (\sigma - \sigma_{\mathrm{i}})^m , \tag{8.31}$$

where ϕ is the orientation factor. Therefore, if $\dot{\varepsilon}$ is measured as a function of σ, the internal stress σ_{i} will be obtained. Most of the techniques described above for the separate determination of internal stress are based on this idea, and efforts have been made to suppress the change in the internal structure during the change in σ or $\dot{\varepsilon}$.

The stress relaxation technique proposed by *Gupta* and *Li* [8.28, 29] obtains σ_{i} from the relation between σ and $\dot{\varepsilon} = -\dot{\sigma}/K$ on the relaxation curve, assuming that the change in internal structure is insignificant because the plastic strain occurring during relaxation is very small. However, as already discussed, recovery is inevitably introduced at high temperatures even for the stress relaxation technique. Even when the strain rate is suddenly changed to suppress the recovery effect, if some effective stress is present the discontinuous change is in the apparent elastic strain rate and not in the plastic strain rate, as will be described in Sect. 8.3.6. Therefore, it is difficult to know the flow stress at the new plastic strain rate without an accompanying change in internal structure. Then, why not change the applied stress suddenly during creep deformation and measure the strain rate immediately after the stress change?

If the effective stress is absent, a stress increase induces an instantaneous plastic strain, as will be described in Sect. 8.3.6, and the internal structure will

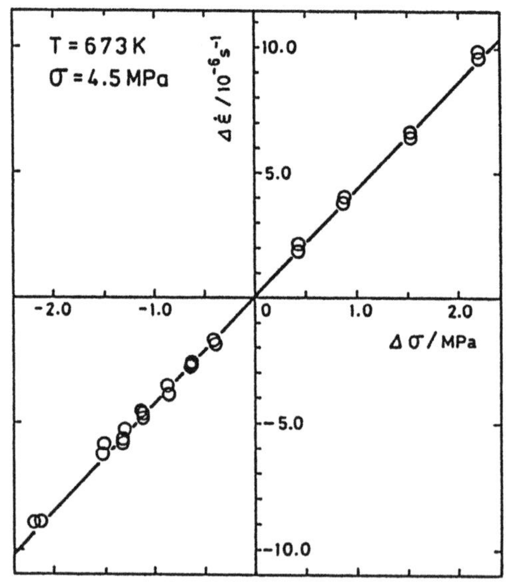

Fig. 8.19. Relation between $\Delta\sigma$ and $\Delta\dot{\varepsilon}$ obtained by applying the stress change technique to steady-state creep of an Al-5.7 at.% Mg alloy [8.30]

change. If some effective stress is present, however, this technique is reasonable, because no instantaneous plastic strain occurs.

Yoshinaga et al. [8.30] measured the strain rate change $\Delta\dot{\varepsilon}$ as a function of stress change $\Delta\sigma$, which was produced during creep deformation of an Al-5.7 at.% Mg alloy (Fig. 8.19). Since a good linear relationship holds between the two variables and thus $m = 1$ in this case, we obtain from (8.31)

$$\frac{\Delta\dot{\varepsilon}}{\Delta\sigma} = \phi^2 \varrho_m b B \ , \tag{8.32}$$

$$\therefore \sigma_i = \sigma - \dot{\varepsilon}\frac{\Delta\sigma}{\Delta\dot{\varepsilon}} \ . \tag{8.33}$$

By using this equation, we can precisely estimate the internal stress.

8.3.5 Fundamental Problems in Internal Stress Measurement

All of the techniques described here are based on the assumption that if the internal structure is the same, the internal stress does not change with a change of applied stress. However, this assumption, though seemingly natural, is actually problematic.

Consider the Orowan equation (8.17) described in Sect. 8.2.2. It is certain that the velocities of simultaneously moving dislocations are different from each other, because the τ_i's on them are not the same and the effective stresses ($\tau_e = \tau - \tau_i$) are different. This is why the average velocity $\bar{v}$ is used in (8.17). Now, a general form

$$v = f(\tau_e) \tag{8.34}$$

is used and the average effective stress corresponding to the average velocity $\bar{v}$ is defined as

$$\bar{\tau}_e = f^{-1}(\bar{v}) \ , \tag{8.35}$$

where $f^{-1}(\bar{v})$ is the inverse function of $f(\bar{\tau}_e)$. Further, let us define the average internal stress as

$$\bar{\tau}_i = \tau - \bar{\tau}_e$$

so that a sum rule holds between the stresses. Thus defined $\bar{\tau}_e$ is called the average effective stress, and $\bar{\tau}_i$ the average internal stress. These average stresses are not the arithmetic average of τ_e's or τ_i's acting on individual dislocations, except in the special case where $f(\tau_e)$ is a linear function, i.e. $m = 1$ in (8.20). Sometimes a technical term, effective internal stress, is used for $\bar{\tau}_i$, [8.31], however, it sounds funny to call $\bar{\tau}_e$ the effective effective stress. In this text the term "average" is used. "Average" means that the stress corresponds to the average velocity of the dislocations.

Consider a particular dislocation moving viscously in a particular place. The internal stress acting on the dislocation is determined by the details of the internal structure at that moment and is independent of the applied stress. The stress that changes when the applied stress is changed is only the effective stress. However, since the average internal stress determined macroscopically is that corresponding to the average velocity as described above, it is not guaranteed to be independent of applied stress. The previously described internal stress is this unguaranteed internal stress. Li [8.31] was the first to conduct a suggestive study about this problem.

The average dislocation velocity described above is the average of velocities of many dislocations spatially distributed in a specimen. However, *Li* and *Chen* et al. [8.32] treated the average velocity as a time average of velocities of a dislocation moving in a periodically fluctuated internal stress field, and calculated how the average internal stress corresponding to this time average velocity changes with the applied stress. Figure 8.20 is *Li*'s result, showing that when a dislocation moves in a sinusoidally fluctuated internal stress field with an amplitude τ_i^0 (see the inset in Fig. 8.20), the average internal stress decreases as the applied stress increases. The parameter m in the figure is the stress exponent[1].

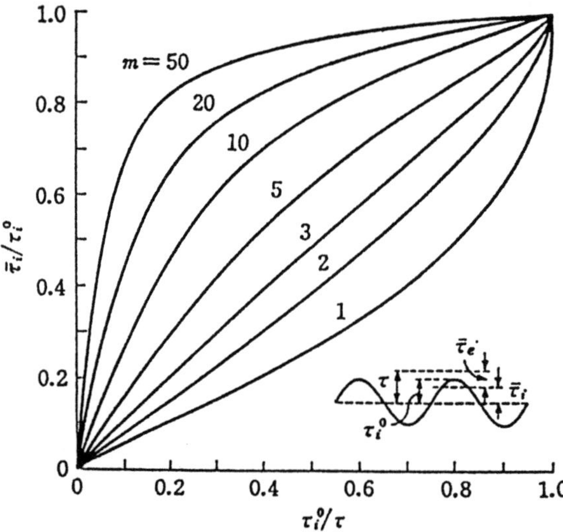

Fig. 8.20. Dependence on applied stress of average internal stress calculated by *Li* [8.31] from the time-average dislocation velocity

---

[1] When $\tau < \tau_i^0$, the dislocation cannot move in a full period and $\bar{v} = 0$. As τ equal to τ_i^0 is the upper limiting stress for $\bar{v} = 0$, then $\bar{\tau}_i = \tau_i^0$. As the applied stress increases further, the effect of the internal stress on dislocation motion gradually decreases and when $\tau_i^0/\tau \to 0$, $\bar{\tau}_i \to 0$. In the range $0 < \tau_i^0/\tau < 1$, the average internal stress is given by [8.31]

$$\frac{\bar{\tau}_i}{\tau_i^0} = \frac{\tau}{\tau_i^0} \left(1 - \frac{1}{x[P_{m-1}(x)]^{1/m}} \right) ,$$

where

$$x = \frac{1}{\sqrt{1 - (\tau_i^0/\tau)^2}}$$

and $P_{m-1}(x)$ is the Legendre function of $(m-1)$th order.

146

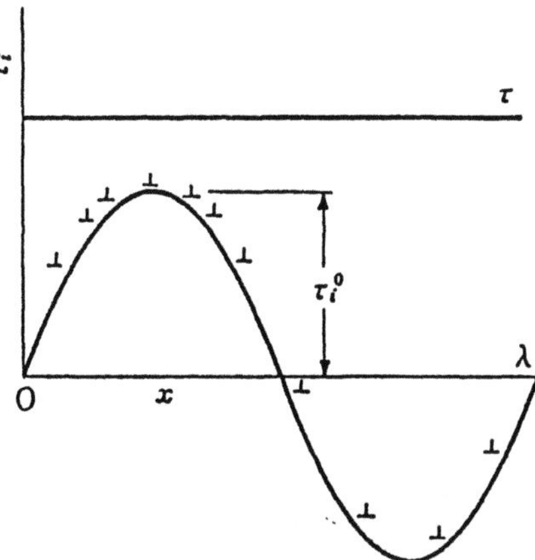

Fig. 8.21. Dislocation distribution in $\tau_i - x$ space

Since the flow stress changes when the temperature or strain rate is changed, any technique devised for internal stress measurement will introduce some change in the average internal stress on the measurement.

The actually required average internal stress is not the internal stress corresponding to the time-average dislocation velocity studied by *Li*, but the internal stress corresponding to the space-average dislocation velocity at an instant. However, according to *Yoshinaga* [8.33], the space average agrees with the time average in a steady state.

Assuming that every moving dislocation experiences the same internal stress fluctuation, and denoting the distance moved by a dislocation by x, spatially distributed dislocations can be superposed on a $\tau_i - x$ space (Fig. 8.21). If τ_i is a periodic function of x, the dislocations can be distributed in a single period, as described by *Ahlquist* and *Nix* [8.26]. Denoting the dislocation density in a region from x to $x + dx$ by $n(x)dx$,

$$\frac{\partial n}{\partial t} = -\text{div} (vn) = 0 \tag{8.36}$$

in a steady state, where the dislocation distribution is independent of time. Therefore,

$$n(x) = \frac{\alpha}{v(x)} \, , \tag{8.37a}$$

where α is a constant independent of x. This means that the density $n(x)dx$ in a place in the $\tau_i - x$ space is proportional to the time for the dislocations to pass that place. Since the total mobile dislocation density is given by

$$\varrho_\text{m} = \int_0^\lambda n(x)dx \, , \tag{8.38}$$

where λ is the period, the proportionality constant α is given by

$$\alpha = \varrho_\text{m} \left(\int_0^\lambda \frac{1}{v(x)} dx \right)^{-1} . \tag{8.39}$$

If the dislocations are distributed as above, the space average of dislocation velocities is expressed by

$$\bar{v} = \frac{1}{\varrho_{\mathrm{m}}} \int_0^\lambda v(x)n(x)dx = \lambda \left(\int_0^\lambda \frac{1}{v(x)}dx \right)^{-1} , \tag{8.40}$$

where

$$t = \int_0^\lambda \frac{1}{v(x)}dx \tag{8.41}$$

is the time for a dislocation to pass a period of the internal stress field. From this, the space average given by (8.40) is known to agree with the time average studied by *Li*.

Since the assumption of the same fluctuation of internal stress is unrealistic, the argument described above may be a very rough approximation. Thus the obtained conclusion may be quantitatively incorrect but qualitatively correct in that the average internal stress changes along with the applied stress even if the internal structure is the same.

The above argument is concerned with the steady state. What is the internal stress like in a non-steady-state immediately after the applied stress has been suddenly reduced? Such an experiment is conducted to suppress the change in internal structure and measure the internal stress during high-temperature deformation. Let us consider this problem by assuming, in the same way as described for the steady state, that the function $\tau_{\mathrm{i}}(x)$ is the same for every dislocation.

From (8.37a, 39, 40), the distribution function of dislocations in a steady state immediately before the stress change is given by

$$n(x) = \frac{\varrho_{\mathrm{m}}\bar{v}}{\lambda v(x)} . \tag{8.37b}$$

When the applied stress is changed so rapidly from τ to τ' during this steady-state deformation that the dislocation distribution $n(x)$ is not changed significantly, the average internal stress will be changed from

$$\bar{\tau}_{\mathrm{i}} = \tau - f^{-1}(\bar{v})$$

to

$$\bar{\tau}_{\mathrm{i}}' = \tau' - f^{-1}(\bar{v}') ,$$

where $\bar{v}$ is given by (8.40) and, from (8.37b), $\bar{v}'$ is given by

$$\begin{aligned}
\bar{v}' &= \frac{1}{\varrho_{\mathrm{m}}} \int_0^\lambda v'(x)n(x)dx \\
&= \frac{\bar{v}}{\lambda} \int_0^\lambda \frac{v'(x)}{v(x)}dx = \frac{\bar{v}}{\lambda} \int_0^\lambda \frac{f(\tau' - \tau_{\mathrm{i}}(x))}{f(\tau - \tau_{\mathrm{i}}(x))}dx .
\end{aligned} \tag{8.42}$$

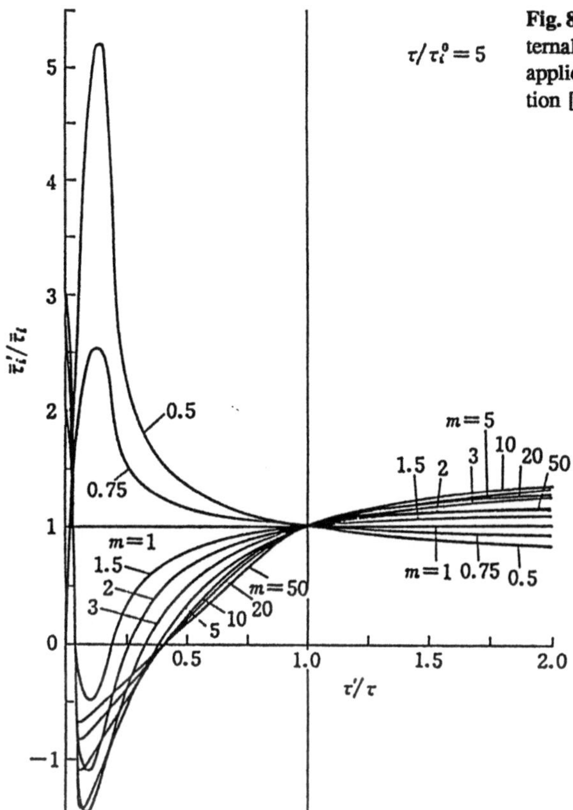

Fig. 8.22. Calculated change in average internal stress caused by a sudden change of applied stress during steady-state deformation [8.34]

When $v = f(\tau_e) = B(\tau - \tau_i)^m$ and $m = 1$,

$$\bar{v}' = \frac{\bar{v}}{\lambda} \int_0^\lambda \frac{\tau' - \tau_i(x)}{\tau - \tau_i(x)} dx = \bar{v} + B(\tau' - \tau) .$$

(8.43)

Accordingly,

$$\bar{\tau}_i' = \tau' - \frac{\bar{v}'}{B} = \tau - \frac{\bar{v}}{B} = \bar{\tau}_i ,$$

(8.44)

which shows that the average internal stress does not change when the applied stress is changed. On the other hand, when $m \neq 1$, the average internal stress will change.

Abe et al. [8.34] calculated how the average stress change depends on the value of m, on the same assumption of sinusoidal fluctuation of internal stress field as that of *Li*. Figure 8.22 is an example of their results, where the applied stress is changed from a steady-state deformation under the applied stress τ five times as high as the internal stress amplitude τ_i^0 to various levels of τ'. As seen in this figure, the averge internal stress changes more as m deviates further from 1.

The above argument is restricted to the case where the dislocation distribution immediately before the stress change is in a steady state. When $m = 1$, however, it is easy to verify that the average internal stress does not change with a change of applied stress for any dislocation distribution and any fluctuation of the internal stress field [8.35]. The experimental results shown in Fig. 8.19 show that this theoretical prediction is correct.

As described above, while the stress change test is a highly reliable technique for measuring internal stress during deformation when $m = 1$, it is most probable that any technique proposed so far gives a larger error as m deviates away from 1.

8.3.6 Techniques to Determine Whether the Effective Stress Is Appreciable or Negligible

As described in the preceding section, when $m \neq 1$, some error is inevitably included in the internal stress measurement. On the other hand, according to the stress dip and extrapolation techniques, the effective stress in pure aluminum is negligible, if any. However, it was shown that the extrapolation technique has a problem regarding experimental sensitivity. Thus some more reliable techniques have been proposed, which do not measure the magnitude of internal stress but determine whether the effective stress is appreciable or negligible.

As described in Sect. 8.2.2, the behavior of dislocation motion is different depending on whether the effective stress exists or not. Further, as will be described in this section, a difference will also be observed in the rate of change of flow stress immediately after the strain rate change. From these differences, the existence of effective stress can be proved experimentally. The following three techniques have been proposed so far.

(a) **Stress Change Technique.** When the effective stress is so negligibly small compared with the flow stress that practically it can be considered absent, a dislocation moves very fast once it surmounts a peak of internal stress. Accordingly, when the applied stress is suddenly increased during deformation, plastic deformation should occur instantaneously until the flow stress increases by work hardening up to the new increased applied stress. On the other hand, when a significant effective stress exists, as long as the applied stress does not rise above the maximum of a thermally surmountable resistance, a dislocation moves viscously. Accordingly, no instantaneous plastic strain should occur, and only an increase in strain rate should be observed. With this in view, *Oikawa* et al. [8.36–38] proposed a technique to determine the presence or absence of effective stress by examining whether instantaneous plastic strain occurs on a sudden increase of applied stress during creep deformation. Their results are shown in Figs. 8.23 and 8.24, where $\Delta\sigma$ is the change of applied stress, Δl_s^+ the instantaneous elongation of specimen for stress increase, Δl_s^- the instantaneous contraction for stress decrease, and $|\Delta\varepsilon_a| = |\Delta l_s|/l_s$. Figure 8.23 is the result for highly solution-hardened alloys. The absolute values of elongation and contraction agree with each other for the same $|\Delta\sigma|$, which shows that the instantaneous strain is only elastic and

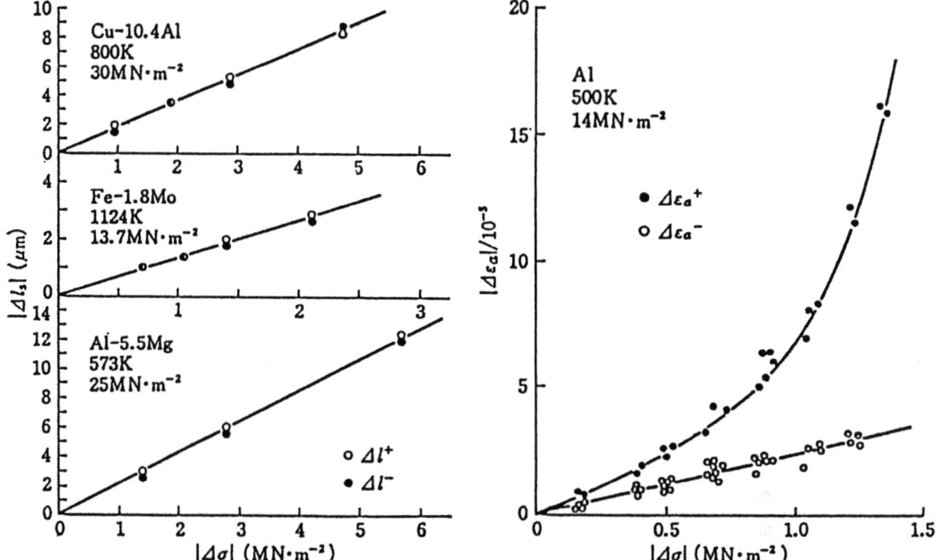

Fig. 8.23. Relation between stress change $|\Delta\sigma|$ and deformation $|\Delta l|$ obtained experimentally by *Oikawa* [8.37]. The solute concentration is in at.%

Fig. 8.24. Instantaneous strain in pure aluminum induced by stress change [8.37]

does not include a plastic component. On the other hand, the result obtained for pure aluminum shown in Fig. 8.24 shows that the absolute value is clearly different between elongation and contraction, which means that instantaneous plastic strain does occur. From these results it is known that the effective stress is not negligible in solution-hardened alloys, but negligible in pure aluminum. This conclusion agrees with that obtained by the stress dip and extrapolation techniques described in Sect. 8.3.3.

b) Strain Rate Change Technique. This technique was proposed by *Yoshinaga* et al. [8.39] to determine the existence of effective stress by measuring the changing rate of flow stress immediately after the strain rate is suddenly changed during deformation.

When the effective and internal stresses coexist, the change in flow stress is given by

$$d\sigma = d\sigma_e + d\sigma_i , \qquad (8.45)$$

where $d\sigma_i$ is determined by the change in internal structure such as dislocation density, and $d\sigma_e$ depends not only on the internal structure but also on the dislocation velocity.

If the applied load is not constant, the displacement speed of the machine cross-head includes the apparent elastic deformation rate described in Sect. 8.3.3

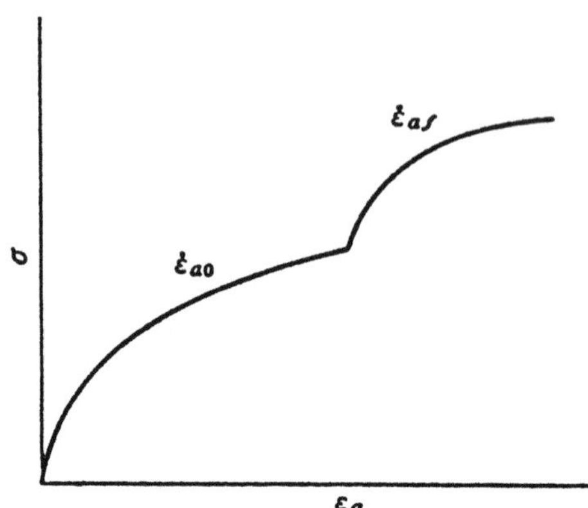

Fig. 8.25. Schematic representation of stress–strain curve obtained by strain-rate change

and is not the same as the plastic deformation rate, i.e., a change in cross-head speed corresponds to the change in apparent strain rate $\dot{\varepsilon}_a$ and does not correspond to the change in plastic strain rate $\dot{\varepsilon}$.

Even when $\dot{\varepsilon}_a$ is suddenly changed, the applied stress can change only continuously, as shown in Fig. 8.25, because of the elastic deformation of the machine assembly. Since the plastic strain rate is finite, the internal structure will also change only continuously. If the changes in applied stress and internal structure are both continuous in this way, the change in plastic strain rate must be continuous. In other words, even when $\dot{\varepsilon}_a$ is changed discontinuously, what changes discontinuously in response to this $\dot{\varepsilon}_a$ change is the apparent elastic strain rate $\dot{\varepsilon}_e$; the plastic strain rate can change only continuously. Discriminating the values immediately before and after the sudden change of $\dot{\varepsilon}_a$ by subscripts 0 and f respectively, from this continuity the equality

$$\dot{\varepsilon}_f = \dot{\varepsilon}_0 \tag{8.46}$$

should hold at the point of sudden change. Accordingly, from (8.27b) and (8.29) we obtain

$$\dot{\varepsilon}_{af} - \frac{\dot{\sigma}_f}{K} = \dot{\varepsilon}_{a0} - \frac{\dot{\sigma}_0}{K} ,$$

or, rewriting this,

$$\dot{\varepsilon}_{af} \left[1 - \frac{1}{K} \left(\frac{d\sigma}{d\varepsilon_a} \right)_f \right] = \dot{\varepsilon}_{a0} \left[1 - \frac{1}{K} \left(\frac{d\sigma}{d\varepsilon_a} \right)_0 \right] .$$

Therefore, the relation

$$\left(\frac{d\sigma}{d\varepsilon_a} \right)_f = K \left\{ 1 - \frac{\dot{\varepsilon}_{a0}}{\dot{\varepsilon}_{af}} \left[1 - \frac{1}{K} \left(\frac{d\sigma}{d\varepsilon_a} \right)_0 \right] \right\} \tag{8.47}$$

will hold between the two slopes shown in Fig. 8.25 on the lower and higher rate sides at that point.

152

Since usually $K \gg (d\sigma/d\varepsilon_a)_0$, (8.47) can be simplified to

$$\left(\frac{d\sigma}{d\varepsilon_a}\right)_f = K\left(1 - \frac{\dot{\varepsilon}_{a0}}{\dot{\varepsilon}_{af}}\right) . \tag{8.48a}$$

This equation can be expressed by the changing rate in flow stress as

$$\dot{\sigma}_f = K(\dot{\varepsilon}_{af} - \dot{\varepsilon}_{a0}) . \tag{8.48b}$$

On the other hand, when the effective stress is negligible,

$$d\sigma = d\sigma_i , \tag{8.49}$$

and the change in internal stress may be given by the difference between its increase $hd\varepsilon$ due to plastic deformation and its decrease rdt due to the annealing for the time $dt\,(= d\varepsilon/\dot{\varepsilon})$, where h is the pure work-hardening rate without recovery effect and r is the pure recovery rate without work-hardening effect. Then,

$$d\sigma = hd\varepsilon - rdt . \tag{8.50}$$

This equation is the same as that once suggested by *Bailey* [8.40] and formulated by *Orowan* [8.41]. In short, (8.50), which is known as the *Bailey-Orowan equation*, holds only when the effective stress is negligible.

Removing $\dot{\sigma}$ from (8.27b, 29 and 50) we obtain

$$\dot{\varepsilon} = \frac{K\dot{\varepsilon}_a + r}{K + h} . \tag{8.51}$$

From this, it is known that when the effective stress is negligible, the plastic strain rate $\dot{\varepsilon}$ will change discontinuously along with $\dot{\varepsilon}_a$. The difference arising from the presence or absence of the effective stress is also reflected in the changing rate of flow stress.

Removing $d\varepsilon$ and $d\varepsilon_e$ from (8.27a, 29 and 50) and rearranging, we obtain

$$\frac{d\sigma}{d\varepsilon_a} = \frac{Kh}{K + h}\left(1 - \frac{r}{h\dot{\varepsilon}_a}\right) . \tag{8.52}$$

As h and r are not considered to change discontinuously, a relation

$$\left(\frac{d\sigma}{d\varepsilon_a}\right)_f = \frac{Kh}{K + h}\left\{1 - \left[1 - \frac{K + h}{Kh}\left(\frac{d\sigma}{d\varepsilon_a}\right)_0\right]\frac{\dot{\varepsilon}_{a0}}{\dot{\varepsilon}_{af}}\right\} \tag{8.53}$$

may hold at the point of sudden change. Usually $Kh/(K + h) \gg (d\sigma/d\varepsilon_a)_0$, and (8.53) can be simplified as

$$\left(\frac{d\sigma}{d\varepsilon_a}\right)_f = \frac{Kh}{K + h}\left(1 - \frac{\dot{\varepsilon}_{a0}}{\dot{\varepsilon}_{af}}\right) , \quad \text{or} \tag{8.54a}$$

$$\dot{\sigma}_f = \frac{Kh}{K + h}(\dot{\varepsilon}_{af} - \dot{\varepsilon}_{a0}) . \tag{8.54b}$$

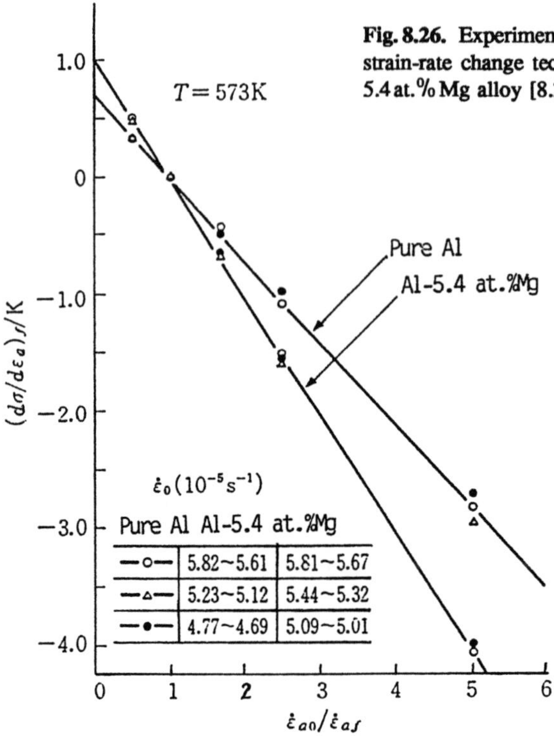

Fig. 8.26. Experimental results obtained by applying the strain-rate change technique to pure aluminum and an Al-5.4 at.% Mg alloy [8.39]

From the above consideration, it is known that the changing rate of flow stress is linear with the ratio of the apparent strain rates (8.48a, 54a) or the difference between them (8.48b, 54b) irrespective of the presence or absence of the effective stress, but the proportionality coefficient depends on the presence or absence of the effective stress. When the effective stress is present, the coefficient is K, and when absent, $Kh/(K + h) < K$.

Figure 8.26 is an example obtained by *Yoshinaga* et al. [8.39] showing that the effective stress is not negligible in a solution-hardened Al-5.4 at.% Mg alloy, but negligible in pure aluminum. This conclusion agrees with that obtained by *Oikawa* et al. using the stress-change test.

c) Stress Relaxation Technique. The stress relaxation technique was proposed by *Abe* et al. [8.42] prior to the techniques of stress change and strain-rate change. As will be described below, this is nothing other than a kind of strain rate change test, but practically very useful because it is very simple.

When the cross-head is arrested during deformation, we obtain a stress relaxation curve as shown in Fig. 8.11. In the present case, however, it is not necessary to conduct such a long-term relaxation as the figure shows; all that we have to do is measure the relaxation rate $\dot{\sigma}$ at the very beginning of the relaxation. The necessary time is usually less than 1s.

To arrest the cross-head is the same as to make $\dot{\varepsilon}_{af} = 0$ for the strain rate change test. Accordingly, the initial relaxation rate $\dot{\sigma}_f$ is given by

$$\dot{\sigma}_f = -K\dot{\varepsilon}_{a0} \tag{8.55}$$

in the presence of effective stress, whereas

$$\dot{\sigma}_f = -\frac{Kh}{K+h}\dot{\varepsilon}_{a0} \tag{8.56}$$

in the absence of effective stress. Since the plastic strain rate is generally related to the apparent strain rate by

$$\dot{\varepsilon} = \dot{\varepsilon}_a\left[1 - \frac{1}{K}\left(\frac{d\sigma}{d\varepsilon_a}\right)\right], \tag{8.57}$$

and since usually $K \gg (d\sigma/d\varepsilon_a)_0$,

$$\dot{\varepsilon}_{a0} = \dot{\varepsilon}_0 . \tag{8.58}$$

Further, the plastic strain rate during relaxation is obtained from the stress relaxation rate by using (8.30). When the effective stress is present, the plastic strain rates before and after the start of relaxation are equal to each other (8.46) at the starting point of relaxation, while when the effective stress is absent[2],

$$\dot{\varepsilon}_f = \frac{h}{K+h}\dot{\varepsilon}_0 < \dot{\varepsilon}_0 . \tag{8.59}$$

Figure 8.27 is a result obtained by *Abe* et al. [8.42] using this technique, and shows that the effective stress is negligible in pure vanadium, but not negligible in a V-5.0 at.% Fe alloy.

As described above, the stress relaxation test is a strain rate change test conducted with only one value of $\dot{\varepsilon}_{af}$. Therefore this technique has the following problems.

Since the strain rate change technique including stress relaxation is used to judge whether the plastic strain rate changes continuously or discontinuously on a discontinuous change of $\dot{\varepsilon}_a$ by the difference in the changing rate of flow stress, a stress change $\delta\sigma$ which is not infinitesimal is needed to measure the changing rate. Therefore, even when the effective stress is present, some change in plastic strain rate, $\delta\dot{\varepsilon}$, may occur. If $\delta\dot{\varepsilon}$ were so large as to no longer be negligible compared with $\dot{\varepsilon}_0$, observations would indicate that the plastic strain rate changed discontinuously and we would reach the wrong conclusion of no effective stress.

---

[2] *Alden* [8.43] derived a relation $\dot{\varepsilon}_f/\dot{\varepsilon}_0 = r/(\dot{\sigma}_0 + r)$ on the wrong assumption of $\dot{\varepsilon}_f = r/h$. Correctly, the binding condition of $\dot{\varepsilon}_{af} = 0$ for the relaxation test and (8.51) lead to $\dot{\varepsilon}_f = -\dot{\sigma}_f/K = r/(h+K)$. Therefore, $\dot{\varepsilon}_f/\dot{\varepsilon}_0 = [r/(\dot{\sigma}_0 + r)][h/(h + K)]$. However, since $K \cong 0.015E$ in the experiment conducted by *Alden* and h is as large as $0.1E$–E, as will be described later, $h/(h + K) \cong 1$. In general, if $h \gg K$, *Alden*'s equation is correct. He stated that $\dot{\varepsilon}_f/\dot{\varepsilon}_0 = 1$ in a steady state of $\dot{\sigma}_0 = 0$. This is also correct only when $h \gg K$.

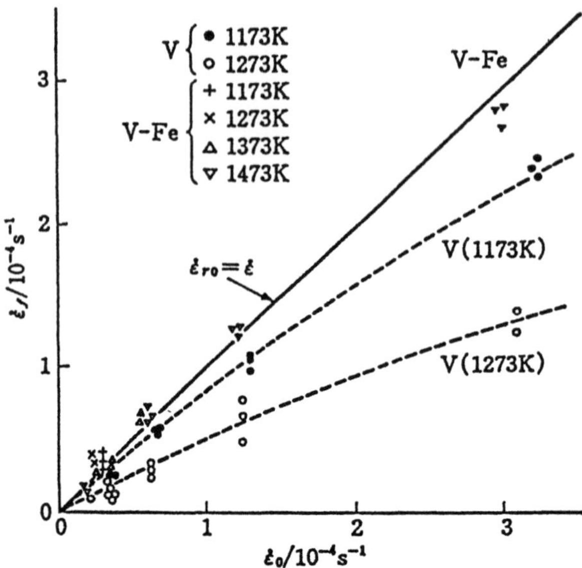

Fig. 8.27. Experimental results obtained by applying the stress relaxation technique to pure vanadium and a V-5.0 at.% Fe alloy [8.42]

Since the relative error $\delta\dot{\varepsilon}/\dot{\varepsilon}_0$ is considered to be proportional to the relative change in effective stress $\delta\sigma/\sigma_e$ [8.39], when $\dot{\varepsilon}_{af} > \dot{\varepsilon}_{a0}$, we find $\delta\sigma > 0$, and the error becomes positive, and when $\dot{\varepsilon}_{af} < \dot{\varepsilon}_{a0}$, we see $\delta\sigma < 0$, and the error becomes negative. Therefore, when data are plotted as shown in Fig. 8.26, the data points will not lie on a straight line passing through point (1,0) [8.39]. Therefore, the strain rate change technique, which uses various $\dot{\varepsilon}_{af}$'s covering both regions of $\dot{\varepsilon}_{af} \gtrless \dot{\varepsilon}_{a0}$, enables us to check whether the judgement is correct or incorrect because we can know the degree of error of $\delta\dot{\varepsilon}/\dot{\varepsilon}_0$ by inspecting the linearity of the $(d\sigma/d\varepsilon_a)_f/K$ versus $\dot{\varepsilon}_{a0}/\dot{\varepsilon}_{af}$ plot. On the other hand, we cannot know the degree of error of $\delta\dot{\varepsilon}/\dot{\varepsilon}_0$ by the relaxation technique. In this respect, the judgement by the relaxation test is less reliable. In order to measure $\dot{\sigma}$ precisely at the point of sudden change, it is recommended to extrapolate the measured values after the sudden change of $\dot{\varepsilon}_a$ to the point of sudden change.

In the above, various techniques that can be used for the separate determination of effective stress and internal stress during deformation have been described. Many of them give an incorrect result. However, the stress dip and extrapolation techniques (Sect. 8.3.3), the stress change technique (Sect. 8.3.4) and the technique to determine the presence or absence of effective stress described in Sect. 8.3.6 are thought to be highly reliable. Even these techniques, however, can give a correct internal stress only when the stress exponent m is unity, and the error increases in principle as m deviates from unity. The error has never been evaluated yet in practical cases, because it depends on how the internal stresses experienced by dislocations are distributed.

9. High-Temperature Deformation Mechanism in Metals and Alloys

In this chapter, the high-temperature deformation mechanisms operating in pure metals and single-phase alloys are described. As obstacles to dislocation motion at high temperatures, jogs in screw dislocations and interdislocation interaction are considered in pure metals, and in addition, the interaction between dislocations and solute atoms in alloys. The resistance of these obstacles of course exists also at low temperatures. However, the effect of diffusion on the resistance becomes severe at high temperatures. The diffusion effect is the subject of this chapter.

9.1 High-Temperature Deformation Mechanism in Pure Metals

The flow stress of pure metals without exception decreases with a rise of temperature and increases with an increase of strain rate. Fig. 9.1 presents the stress-strain curves of polycrystals of aluminum with fcc structure and of vanadium with bcc structure[1] at various temperatures. As indicated in the figure, the strain rate was changed by a factor of 10 during deformation. The temperature and strain-rate dependences of flow stress are clearly seen in the figure. As described in Sect. 8.3.2, these dependences can qualitatively be explained from the point of view from the effective stress or the internal stress. In the following, the important deformation mechanisms proposed so far and their validity will be described.

9.1.1 Jog-Drag Theory

The stress for dragging jogs in screw dislocations is a possible origin of the effective stress at high temperatures, because the activation energy for jog dragging is fairly high. The motion of a jog with a screw dislocation is non-conservative and needs the formation of a vacancy or self-interstitial. However, because the formation energy is much higher for an interstitial than for a vancancy, actually at high temperatures an interstitial-forming jog is considered to climb by absorbing vacancies.

---

[1] The metals of fcc or bcc structure satisfy the von Mises criterion [9.2] of "more than 5 independent deformation systems". Therefore, the deformation behavior of polycrystals is not much different from that of single crystals, especially in the highly deformed state

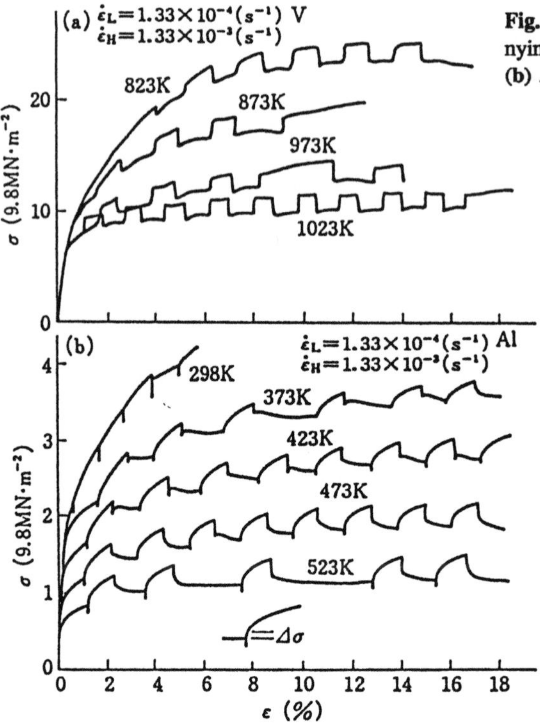

Fig. 9.1a,b. Change in flow stress accompanying strain-rate change [9.1]. (a) Vanadium (b) Aluminum

Hirsch and *Warrington* [9.3] deformed aluminum and copper at high temperatures and again at 273 K, and obtained the ratio r_H of the flow stress at high temperatures at the point of unloading to the yield stress at 273 K. They confirmed that r_H decreases rapidly in the high temperature region in a similar manner to Fig. 8.9. During cooling to 273 K, a significant recovery probably occurred in the internal structure of the specimens. However, since the recovery effect, if it exists, will make the yield stress decrease, the observed decrease in r_H at high temperatures can be an underestimation but never an overestimation. In other words, the fact that a decrease in r_H is recognized in the high temperature region even by their method means that the flow stress, if it is measured under the same internal structure condition, is sure to decrease at high temperatures (presence of effective stress). A problem occurs if the yield stress measured at 273 K does not really include a significant work hardening due to the deformation at 273 K. In this respect they pointed out that the yield stress was measured at a plastic strain much less than 0.2%, and assumed that such a small plastic strain would not introduce a significant work hardening. Based on this result, they proposed a jog-drag mechanism, and theoretically derived a rate equation

$$\dot{\gamma} = f N_a l_j b^2 \nu^* z \exp\left(\frac{S_{SD}}{k_B}\right) \exp\left(-\frac{Q_{SD} - V_a \tau_e}{k_B T}\right) . \tag{9.1}$$

Here, N_a is the number of activation points per unit volume, l_j the mean jog

158

spacing along the dislocation, S_{SD} and Q_{SD} the entropy and the energy for self-diffusion, respectively, V_a the activation volume, and τ_e the effective stress. N_a is given by ϱ_m/l_j, and $(b^2/6)\nu^* z \exp(S_{SD}/k_B) \exp(-Q_{SD}/k_B T)$ is the self-diffusion coefficient, D. Since a force $l_j \tau_e b$ acts on a jog and the jog climbs by b for every activation event, the work done by τ_e for an activation process is $l_j \tau_e bb = V_a \tau_e$, and the activation energy is reduced by this amount. When the vacancy concentration around a jog, c_v, differs from the equilibrium concentration, c_{ve}, a chemical force of $(k_B T/b^2) \ln(c_v/c_{ve})$ [9.4] will also act on the jog. This effect is included in (9.1) as the factor f. When the density of vacancy sinks is sufficiently high and c_v is nearly the same as c_{ve}, the rate of emission of vacancies may control the climb of the jog and then $f = 1$, while when the sink density is low and c_v is very different from c_{ve}, the diffusion of vacancies may control the jog climbing and then $f < 1$, because the chemical force hinders the climbing. Edge dislocations are thought to glide much faster than screw dislocations and the contribution to the total strain may not be much different between edge and screw dislocations. Then the actual strain rate may be doubled.

Hirsch and *Warrington* considered that when the jog is not a single one of height b but a multiple jog of height $n_j b$, the activation volume is reduced by a factor of $1/n_j$, and they showed that all their experimental results are well explained by assuming that $n_j = 2$–5 in aluminum and 1–2 in copper. The conservative motion of a jog along a screw dislocation line should become faster when the extension of the jog is narrower, n_j is smaller, and the temperature is higher. Then, in aluminum of a high stacking fault energy, jogs of a small n_j may be annihilated with the conservative motion and only the jogs of a large n_j resist the screw dislocation motion. The difference in n_j between aluminum and copper corresponds well to the stacking fault energy between them.

In the intermediate temperature region of B of Fig. 8.9, τ_H is almost independent of temperature. This observation can be understood by considering that the energy of thermal vibration is small compared with the activation energy for self-diffusion and jog dragging becomes a athermal process. From this point of view, the critical temperature separating the high and intermediate temperature regions is given as the temperature at which the athermal resistance is equal to τ_e.

From (9.1), τ_e is given as

$$\tau_e = \left[Q_{SD} - k_B T \ln \left(\frac{f N_a l_j b^2 \nu^* z \exp(S_{SD}/k_B)}{\dot{\gamma}} \right) \right] \Big/ V_a$$

and the athermal resistance is given by

$$\tau_i = \frac{U_0}{l_j b^2} = \frac{U_0}{n_j V_a} \, ,$$

where U_0 is the energy per jog for the dislocation to move by b. From the condition $\tau_e = \tau_i$, the critical temperature is expressed as [9.3]

$$T_c = \frac{Q_{SD} - U_0/n_j}{k_B \, \ln[f N_a l_j b^2 \nu^* z \exp(S_{SD}/k_B)/\dot{\gamma}]} \, . \tag{9.2}$$

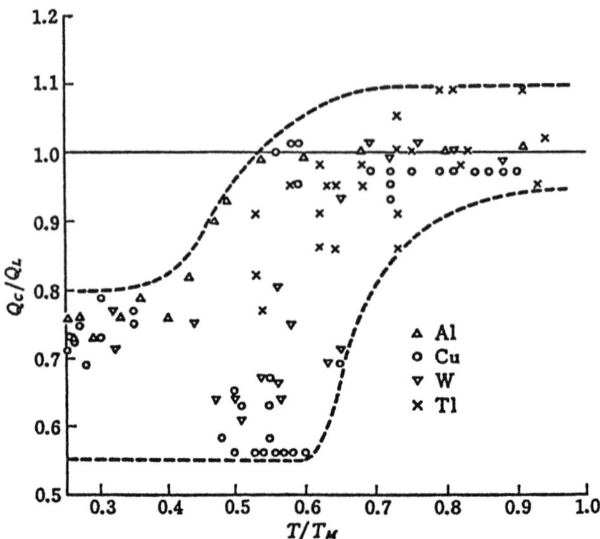

Fig. 9.2. Homologous temperature dependence of activation energy for creep, Q_c, normalized by the activation energy for lattice diffusion, Q_l, (*Evans* and *Knowles* [9.5])

The activation energy for deformation in general depends on the temperature range as shown in Fig. 9.2 [9.5] and it is well known that at sufficiently high temperatures the energy for steady-state creep agrees well with the activation energy for self-diffusion [9.6]. If the internal structure of a metal being deformed in steady state is the same irrespective of temperature for a given applied stress, both the effective stress and the dislocation density should be the same, and thus the agreement of the two energies suggests that the dislocation velocity is proportional to the self-diffusion coefficient. The proportionality agrees with a prediction from the jog-drag theory for the low stress creep where $Q_{SD} \gg V_a \tau_e$.

Since the study by *Garofalo* [9.6], the stress dependence of the steady-state creep rate has been known to be expressed in the form [sinh $(\alpha' \sigma)]^n$ over a wide stress range, where $\alpha' = V_a/k_B T$. This function gives a power law, $\dot{\varepsilon} \propto \sigma^n$, in the lower stress range and a stress dependence $\dot{\varepsilon} \propto \exp(V_a\sigma/k_B T)$ in the higher stress range. Since the theory of *Hirsch* and *Warrington* is not for the steady-state deformation, it does not directly concern this stress dependence. *Nix* and coworkers tried to explain the stress dependence of the steady-state creep rate by jog-drag theory [9.7–9]. After having investigated various aspects, they showed that the density of the vacancy-forming jogs must be not less than the density of vacancy-absorbing jogs in order to explain the experimental facts. This condition is also a premise of the theory of *Hirsch* and *Warrington*.

If the jog density is in thermal equilibrium, the densities of the two kinds of jogs are equal to each other, and the above condition is satisfied. In this case, however, the activation energy for deformation should disagree with the activation energy for self-diffusion, because the jog density depends on temperature.

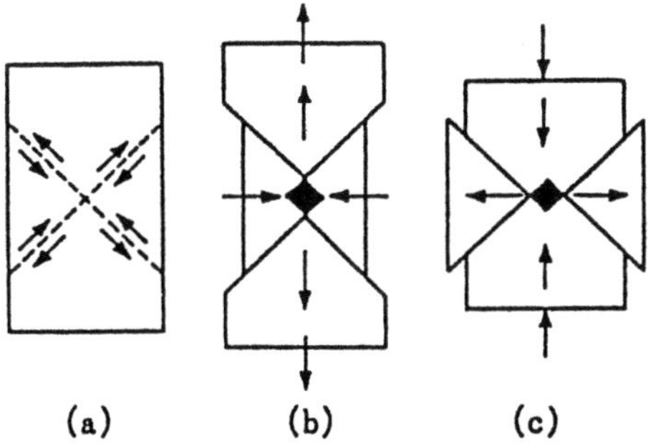

Fig. 9.3a–c. Illustration of the formation of interstitial atoms (or absorption of vacancies) when slip occurs simultaneously on two slip planes intersecting each other [9.12]. (a) Before deformation. (b) After tensile deformation. (c) After compressive deformation. The black region indicates substance superposition

Since the jog density in thermal equilibrium is estimated to be very low [9.7], the majority of jogs are considered to be produced by mutual cutting of dislocations. If this is the case, the above difficulty concerning the activation energy is removed. In this case, however, another difficulty will arise.

As shown by *Cottrell* [9.10], when two or more slip systems are activated simultaneously, the jog produced by mutual cutting of screw dislocations are mostly of vacancy-absorbing type. *Weertman* [9.11] pointed out that the jog-drag theory has a difficulty in this respect. Figure 9.3 is an illustration given by *Weertman* [9.12] showing that when two slip systems operate simultaneously they are forced to form interstitial atoms or absorb vacancies. Naturally, when two slip systems operate sequentially or only one slip system operates, the difficulty does not occur. However, in polycrystals naturally, and even in single crystals, vacancy-absorbing jogs may be produced more often than vacancy-forming jogs in a highly strained state. *Malu* and *Tien* [9.13] theoretically examined the jog-drag theory on the premise that the vacancy-absorbing jogs are more prevalent. However, many of the experimental results do not seem to support the theoretical prediction.

If the jog drag is the rate determining process for deformation, some effective stress required for the drag must exist. Therefore, if it is verified experimentally that the effective stress does not exist or is negligibly small compared with the internal stress, the jog-drag theory will in practice lose its usefulness.

Many experiments have been carried out to separately determine the effective stress and internal stress components in flow stress. However, since the stress dip technique mentioned in Sect. 8.3.3 was proposed, this technique has been mostly used, because the technique is believed most reliable for minimizing change in internal structure during the measurement. Unfortunately, the obtained results are divided into two groups conflicting with each other.

One of the two is represented by the study of *Ahlquist* and *Nix*, [9.14] (Fig. 9.4), indicating that the effective stress certainly exists and that its fractional contribution to flow stress, $\sigma_e/\sigma = 1 - \sigma_i/\sigma$, increases with an increase in flow stress and with a rise of temperature. These results were regarded as strong

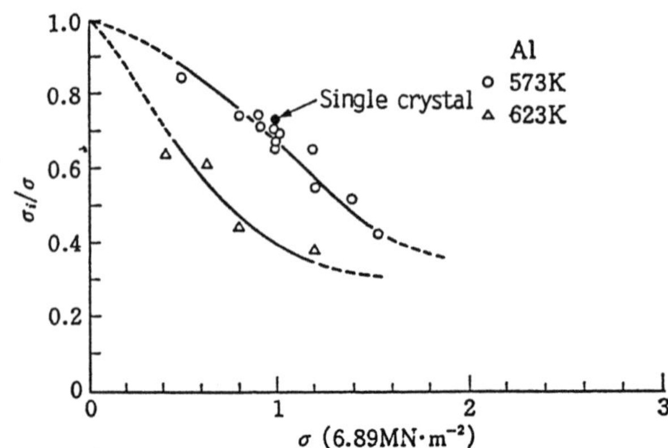

Fig. 9.4. Stress dependence of internal stress in steady-state creep of aluminum [9.14]

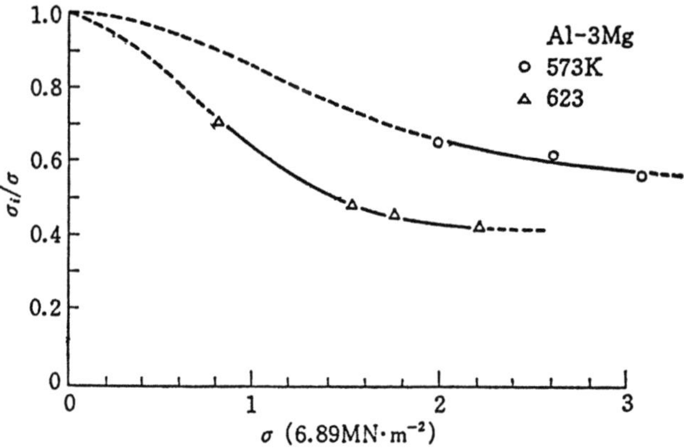

Fig. 9.5. Stress dependence of internal stress in steady-state creep of Al-3 at.% Mg alloy [9.14]

evidence for jog drag. On the other hand, as shown in Fig. 9.5, the reported experimental values of σ_e/σ do not vary greatly between pure aluminum and Al-Mg alloys in spite of the fact that the effective stress for solute-atmosphere drag is considered very high in solution-hardened alloys such as Al-Mg. Such an incomprehensible result throws doubt upon the reliability of the measurement.

The other results contradict the above. Figure 9.6 is a typical example reported by *Davies* et al. [9.15]. When the accuracy of measurement is very high, a short time deformation stagnation is recognized immediately after the stress reduction even if the amount of stress reduction is very small (5% of flow stress). These results show that the effective stress is negligibly small.

As described in Sect. 8.3.3, *Toma* et al. [9.16] showed that the effective stress is negligible, within the experimental error, by using the stress dip and extrapolation technique. It is very probable that in the paper of *Ahlquist* and *Nix* and in many other papers reporting similar results to theirs, the short time deforma-

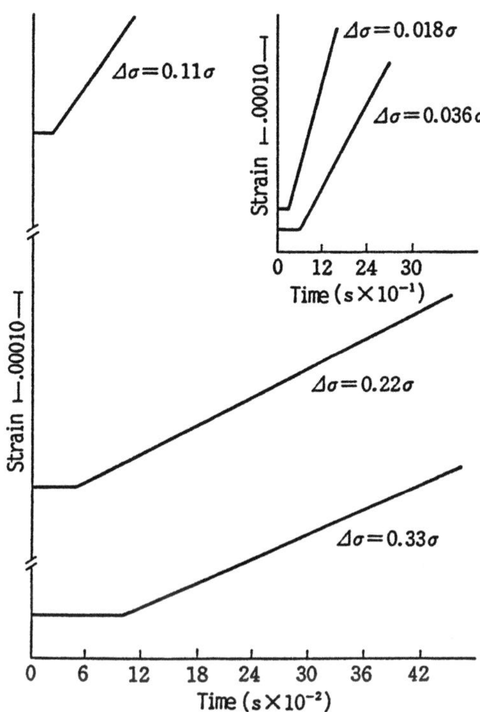

Fig. 9.6. Strain transient after stress reduction during steady-state creep of pure copper at 686 K under 48 MN m^{-2} [9.15]

tion stagnation for small stress reductions was not detected because of a slow reduction rate of the applied stress or a slow response of the recorder.

Since then many observations have been reported of instantaneous plastic strain being produced by a sudden increase of applied stress (Fig. 8.24), the plastic strain rate being changed discontinuously by a sudden change of apparent strain rate (see Fig. 8.26 for the strain rate change technique, and Fig. 8.27 for the stress relaxation technique). These observations show that the effective stress is negligible in pure fcc and bcc metals.

When the effective stress is negligible, from (8.50) the strain rate is given by

$$\dot{\varepsilon} = \frac{\dot{\sigma} + r}{h} \, . \tag{9.3}$$

For a creep test or a steady-state deformation tensile test, $\dot{\sigma} = 0$ and the strain rate becomes

$$\dot{\varepsilon} = \frac{r}{h} \, . \tag{9.4}$$

This equation shows that the deformation is controlled by recovery (sometimes by recrystallization).

9.1.2 Theory of Recovery Control

The idea that the flow stress at high temperature is determined by the competition of work hardening and recovery was proposed by *Bailey* [9.17] and formulated as (8.50) by *Orowan* [9.18], and is called the *recovery control theory*. If the recovery occurs through dislocation climb, the experimental fact that the strain rate is proportional to the self-diffusion coefficient can be explained from (9.4), because the dislocation climb is controlled by diffusion. Much effort has been applied to theoretically deriving the observed stress dependence of strain rate.

Weertman [9.19–21] was the first to do this derivation. He assumed a dislocation pile-up against a sessile dislocation [9.19] produced by a multiple slip or against a dislocation multipole produced by mutual arresting of oppositely signed dislocations on parallel slip planes. Whenever the leading dislocation of the pile-up climbs over the sessile dislocation or a pair annihilation occurs in the multipole, a dislocation loop is emitted from a source to compensate the reduction in back stress acting on the source, because the climb over or the pair annihilation reduces the back stress of the pile-up. On these assumptions, he calculated the strain rate of steady-state creep as a function of the applied stress. In this process, the distance the dislocation needs to climb is much smaller than the glide distance from the source to the obstacle to be climbed. Therefore, the rate determining process for deformation is the dislocation climb, though most of the deformation is carried by the glide motion.

According to this calculation, the strain rate is given by

$$\dot{\gamma} = \frac{2\pi^{1/2}D}{3b^{7/2}M^{1/2}\ln(d_{\mathrm{s}}/4b)} \left(\frac{\tau}{\mu}\right)^{7/2} \left(\frac{\tau\Omega}{k_{\mathrm{B}}T}\right) , \tag{9.5}$$

where M is the density of dislocation sources and d_{s} the spacing of active slip planes. If the source density is assumed to be independent of temperature and stress, this equation shows that the steady-state creep rate is proportional to the 4.5th power of applied stress. This equation was derived by approximating the climbing velocity [proportional to $\exp(\sigma_{\mathrm{n}}\Omega/k_{\mathrm{B}}T) - 1$] as being proportional to $\sigma_{\mathrm{n}}\Omega/k_{\mathrm{B}}T$ under the condition $\sigma_{\mathrm{n}}\Omega/k_{\mathrm{B}}T \ll 1$. Therefore, the theory predicts that the approximation becomes worse as the stress increases and the power law breaks down in the high stress range. This prediction agrees with the experimental observation. Here, σ_{n} is the normal stress aiding the dislocation climb. *Weertman* used as σ_{n} the mutual stress field acting on the oppositely signed dislocations. The difficulty of this theory is the unrealistic assumption of constant M, because the dislocation density is remarkably dependent on the creep stress and thus the source density cannot remain constant.

It is well known that the dislocation structure in pure metals deformed at high temperature is the cell structure. Accordingly, the deformation theory should be constructed on the premise of this cell structure. The cell wall becomes thinner as the temperature rises and the strain rate is lowered, and it finally becomes a sub-boundary (Fig. 9.7).

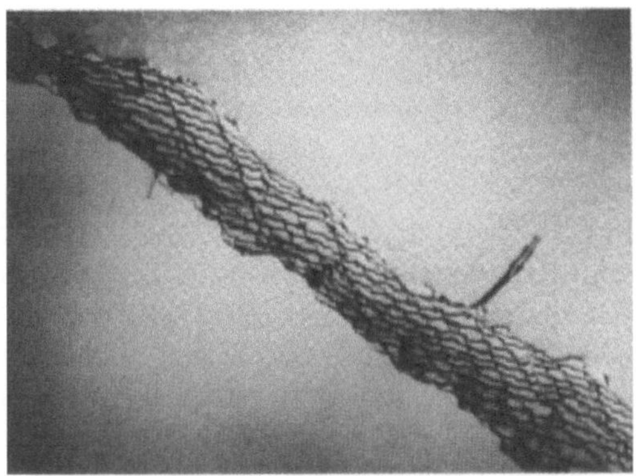

Fig. 9.7. Dislocation structure observed in pure aluminum crept to 11% at 632 K under 2.8 MN m^{-2}, ×30000 (Courtesy of *Horiuchi* and *Otsuka*)

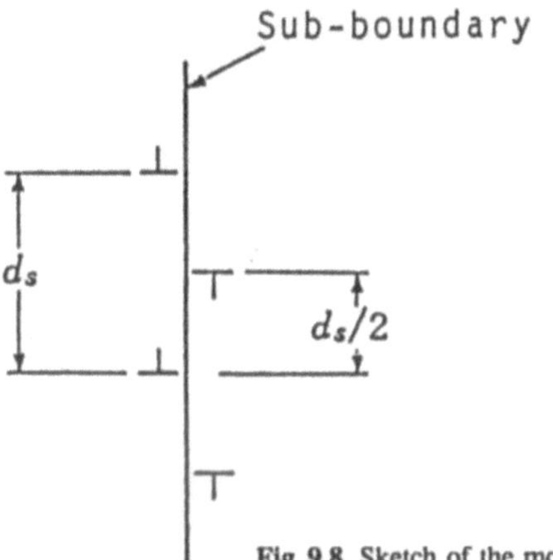

Fig. 9.8. Sketch of the model proposed by *Ivanov* and *Yanushkevich* [9.22]

Ivanov and *Yanushkevich* [9.22] are the first to propose a model considering the existence of sub-boundaries. They assumed that slip occurs with a spacing d_s in subgrains of volume L_s^3 and that the dislocation motion is hindered by the sub-boundary. As shown in Fig. 9.8, once the opposite dislocations blocked on both sides of the sub-boundary are pair-annihilated after climbing about $d_s/4$, new dislocation loops are produced and each of them sweeps an area of about L_s^2. They considered that deformation proceeds by the process. This model is similar to that of *Weertman*, except that no dislocation pile-up is considered, and gives the 3rd-power law of stress, i.e.,

$$\dot{\gamma} \cong 20 \left(\frac{D}{b^2}\right) \left(\frac{\tau}{\mu}\right)^2 \left(\frac{\tau\Omega}{k_B T}\right) . \tag{9.6}$$

In this theory, the applied stress is used as the driving force of dislocation climb. If, instead, the interaction of opposite dislocations is used (as by *Weertman*), this theory would give the 4th power law, because the climbing velocity will increase as the spacing of the dislocations decreases. Further, if dislocations pile-up against the sub-boundary, the stress concentration factor increases in proportion to $L_s\tau$ and if in addition the experimental relation $\tau \propto 1/L_s^{m_s}$ is used, the stress exponent increases by $(1 - 1/m_s)$.

The theory of *Blum* [9.23] belongs to this kind of model and the concept is not much different from the above. How the stress exponent varies according to the model used is described in detail in the work of *Poilier* [9.24].

In the above theories, the sub-boundary is considered merely as a dislocations sink. Conversely, it can act as a dislocation source when the dislocation network constructing the sub-boundary is coarsened by recovery [9.25, 26]. However, there exist dislocations in subgrains, though not many, and they are regarded as constructing a network structure which is coarser than the network constructing the sub-boundary [9.27]. Then it is natural to consider that such a coarse network should act as the source of dislocations.

The climb motion of dislocations is required for the network to coarsen, and thus the coarsening is also controlled by diffusion. When dislocation links in a grown network bow out and glide dislocations are emitted, they are incorporated into another part of the network to reduce the average mesh size of the network. The growing rate of links may correspond to the recovery rate r and the mesh size reduction rate may correspond to the work-hardening rate $h\dot{\gamma}$. When they balance, a steady state is realized.

There is a relation

$$\varrho = \frac{1}{l_1^2} \tag{9.7}$$

between the mean link length l_1 and the dislocation density ϱ. In order for the link length to grow by δl_1, the links must climb about δl_1, and for the climb it is required that vacancies $(\varrho/b)(\delta l_1/b)$ in number are emitted or absorbed. From (9.7), the decrease in dislocation density due to this growth is given by

$$-\delta\varrho = \frac{2\delta l_1}{l_1^3} . \tag{9.8}$$

Assuming that the dislocation self-energy is $\mu b^2/2$ per unit length, we obtain for the decrease in dislocation energy per unit volume

$$-\delta E = -\frac{\mu b^2}{2}\delta\varrho = \mu b^2 \frac{\delta l_1}{l_1^3} . \tag{9.9}$$

This means that a work done is

$$-\frac{\delta E b^2}{\varrho \delta l_1} = \frac{\mu b^4}{l_1}$$

per vacancy on emission or absorption. If it is assumed after *Friedel* [9.4] that in the case of $l_1 < 30\,\mu m$ the chemical force arising from the difference in vacancy density from the equilibrium value can be neglected compared with the

dislocation line tension, the climbing velocity of dislocations will be controlled
by the emission or absorption of vacancies, and thus the velocity is given by

$$v_{cl} = \frac{dl_1}{dt} = \frac{D}{k_B T} \left(\frac{\mu b^4}{l_1} \right) \frac{c_j}{b} = \frac{\mu b^3 c_j}{k_B T} \frac{D}{l_1} \, ,$$

(9.10a)

where c_j is the jog density. It is because the effect of pipe diffusion along a
dislocation line is neglected and the vacancy emission is considered here to be
the rate-controlling process that the factor c_j comes into this equation. On the
other hand, in the case of lattice diffusion control, the vacancy density may be
almost uniform along a dislocation link, because the pipe diffusion is much faster
than the lattice diffusion. In this case, the diffusion distance of vacancies is the
spacing between the vacancy emitting links and absorbing links, and thus the
above equation should be modified to [9.5]

$$v_{cl} = 2\pi \frac{\mu b^3}{k_B T} \frac{D}{l_1 \ln(l_1/2b)} \, .$$

(9.10b)

If dislocation links of length l_1 act as dislocation sources, the flow stress is
approximately given by

$$\tau = \frac{\mu b}{l_1} \, .$$

(9.11)

McLean and coworkers [9.27, 28] derived the 3rd power of stress dependence of
recovery rate r from (9.10a) and (9.11) as

$$r = -\frac{\partial \tau}{\partial t} = -\frac{d\tau}{dl_1} \frac{\partial l_1}{\partial t} = \frac{b c_j}{\mu k_B T} D\tau^3 \, .$$

(9.12)

In the same way, the work hardening rate h, which competes with the recovery, is given by

$$h = \frac{\partial \tau}{\partial \gamma} = \frac{d\tau}{dl_1} \frac{\partial l_1}{\partial \gamma} \, .$$

(9.13)

Here,

$$d\gamma = d\varrho s b = dN S b \, ,$$

(9.14)

where $d\varrho$ is the increase in dislocation density corresponding to the increase in
flow stress $d\tau$. It decreases the link length as seen from (9.8). dN is the increase
in dislocation loop number per unit volume. The values of s and S are the glide
distance of a dislocation and the area swept by a dislocation loop, respectively.
When a dislocation loop is stopped after expanding to a radius r_s, $S = \pi r_s^2$ and
the loop length is $2\pi r_s$. Thus, we have following relations:

$$d\varrho = 2\pi r_s dN \, , \quad s = \frac{r_s}{2} = \left(\frac{S}{4\pi} \right)^{1/2} \, .$$

(9.15)

From (9.8, 11, 13 and 14),

$$h = \frac{\mu l_1}{2s} = \frac{\mu^2 b}{2s\tau} .$$ (9.16)

According to this network growth theory, the strain rate is given by substituting (9.12) and (9.16) for r and h in (9.4) as

$$\dot{\gamma} = 2 \left(\frac{\mu b D}{k_{\mathrm{B}} T} \right) c_j \left(\frac{s}{b} \right) \left(\frac{\tau}{\mu} \right)^4 .$$ (9.17)

If the parameter s is independent of stress, this equation predicts that the strain rate is proportional to the 4th power of stress. However, the spacing of obstacles, which determines the glide distance s, is in general dependent on stress. *H.E. Evans* [9.29] assumed $s \cong l_1$ from the idea that the dislocation source is the network in subgrains. In his case, $h = \mu/2$ independent of stress, giving the 3rd power of τ dependence of $\dot{\gamma}$. However, the stress exponent n is in many cases about 4.5 in pure metals, as shown by Weertman's equation (9.5). Thus, *W.J. Evans* [9.26] proposed a model that the dislocation source exists on a sub-boundary and s is about the size of a subgrain L_{s}. As described previously, $\tau \propto 1/L_{\mathrm{s}}^{m_{\mathrm{s}}}$, and thus, $h \propto \tau^{1/m_{\mathrm{s}}-1}$ and $\dot{\gamma} \propto \tau^{4-1/m_{\mathrm{s}}}$. According to the experimental results on copper and aluminum given by *Staker* and *Holt* [9.30] and *Orlová* et al. [9.31], $m_{\mathrm{s}} = 1$ in the higher stress region of $\tau/\mu b \gtrsim 2[\mu m]^{-1}$ and $m_{\mathrm{s}} = 2$ in the lower stress region [9.26]. Therefore, n becomes not more than 3.5 and the model cannot explain the experimental value. Therefore, *W.J. Evans* introduced a threshold stress τ_0 for deformation, and used $(\tau - \tau_0)$ in place of τ. Then the apparent stress exponent n' becomes larger than the true exponent n, because

$$n' = \frac{\partial \ln \dot{\gamma}}{\partial \ln \tau} = \frac{\tau}{\tau - \tau_0} \frac{\partial \ln \dot{\gamma}}{\partial \ln(\tau - \tau_0)} = n \frac{\tau}{\tau - \tau_0} .$$ (9.18)

Thus n' can take various values larger than n, depending on the ratio of τ/τ_0. *W.J. Evans* gave no theoretical evaluation of τ_0. Later, *Caillard* and *Martin* [9.32, 33] found that the sub-boundaries bulge out between triple nodes. *Argon* and *Takeuchi* [9.34] conducted a detailed analysis based on the idea that the long-range internal stress produced by the bulging is the origin of τ_0.

On the other hand, *Parker* and *Wilshire* [9.35] showed that the sub-boundaries do not play an important role, on the basis of their experimental results that the creep rate depends not on the subgrain size but on the dislocation network inside the subgrains. Taking this view, *H.E. Evans* and *Knowles* [9.5] examined in detail the above-described growth mechanism of *Friedel* [9.4] concerning the dislocation network in subgrains, and showed that if s is independent of τ, the equation derived by them can predict many of the experimental creep rates within a factor of 6. As described above, the stress exponent is 4 in this case. Thus, the accuracy of the prediction becomes better than $\pm 56\%$ for the flow stress. However, the reason for the stress independence of s is not clear.

When the applied stress is suddenly reduced during creep deformation, an instantaneous plastic strain occurs in the reverse direction [9.15, 36]. From this, *Davies* et al. [9.15] considered that some dislocation pile-ups are built up even during creep at high temperature. If dislocations are emitted in groups from the links having grown to a length sufficient to act as a source under the applied stress and pile up against barriers after producing an instantaneous plastic strain, it is also possible that n increases through the stress dependences of the number of links acting as dislocation sources and the number of dislocations in a pile-up.

9.1.3 Effect of Inhomogeneity in the Dislocation Structure

The theories described above do not take into account the inhomogeneity in dislocation structures. The link length can vary in an actual dislocation network and accordingly the critical stress for the link to act as a dislocation source (9.11) varies. The stress required to activate a link of length l_1 is given by

$$\tau_y = \frac{\mu b}{l_1} + \tau_i \,, \tag{9.19}$$

where τ_i is the internal stress acting on the link arising from other dislocations. Considering τ_y as the strength of the link and denoting the link density of strength ranging from τ_y to $\tau_y + d\tau_y$ by $M(\tau_y)d\tau_y$, we find the distribution function $M(\tau_y)$ may be of the form shown in Fig. 9.9 for example. This idea was proposed by *McLean* [9.37–39]. When a stress τ is applied to a specimen having such internal structure, the weak links of $\tau_y \leqq \tau$, i.e.,

$$l_1 \geqq \frac{\mu b}{\tau - \tau_i} \tag{9.20}$$

(the part indicated by a light broken curve) all are moved instantaneously and exhausted. The change in internal structure introduced by this movement will change $M(\tau_y)$ in the region $\tau_y > \tau$.

When a stress is applied, an instantaneous plastic strain is induced by the dislocations of $\tau_y < \tau$ and the deformation will stop provided no recovery proceeds afterwards and the applied stress is not increased. However, if recovery proceeds, the peak in the figure will be lowered, as shown by an arrow, the

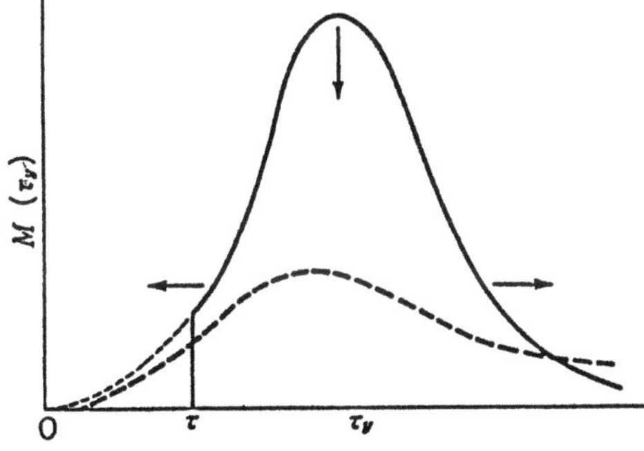

Fig. 9.9. Schematic distribution of link strength

distribution function tending to become the bold broken curve [9.40]. Therefore, the weak links of strengths less than τ are continuously supplied and deformation proceeds without any increase of τ. In this case, the strain increment $\Delta\gamma$ induced by an increase of applied stress $\Delta\tau$ may be given by

$$\Delta\gamma \cong M(\tau)\Delta\tau b\bar{S} . \tag{9.21}$$

Therefore, the work-hardening rate h may be given by

$$h \cong \lim_{\Delta\tau \to 0} \frac{\Delta\tau}{\Delta\gamma} = \frac{1}{M(\tau)b\bar{S}} . \tag{9.22}$$

Here, $\bar{S}$ is the average area swept out by an emitted dislocation loop, and is determined by the dislocation structure and the applied stress. Equation (9.22) shows that h can become very high when $M(\tau)$ is very small. The recovery rate is given by

$$r = \left(\frac{\partial\tau_y}{\partial t}\right)_{\tau_y = \tau} \tag{9.23}$$

If the internal stress due to other dislocations, τ_i, is not included in τ_y, r is given by McLean's equation (9.12). However, the growth rate of links to be considered here is not the average velocity of (9.10) given by *Friedel* but the growth rate of the weakest (longest) link. Further, the contribution of τ_i to τ_y may not be negligible. However, τ_i is also expected to be inversely proportional to l_1 as is the first term on the right-hand side of (9.19) and the growth rate of the weakest link may be approximately proportional to the average growth rate. Accordingly, it is considered that these effects give only a change in the numerical factor, and the temperature and stress dependences of $r(\propto D\tau^3)$ remain unchanged.

Alden [9.41] proposed an idea similar to the idea of strength distribution of dislocation sources: an inhomogeneous distribution of athermal obstacles (other dislocations in the present case) to dislocation motion. As argued by *Kocks* [9.42], because of the inhomogeneous distribution the swept area $\bar{S}$ becomes larger as the applied stress is increased, provided the stress is lower than the critical level for the total area to be swept. *Alden* referred to the ratio of $\bar{S}$ to the total area, β, as the relative free area function. He assumed the stress dependence of β as

$$\beta = \exp\left(-\frac{\tau_0 - \tau}{\tau_V}\right) , \tag{9.24}$$

and compared the theoretical prediction with experimental results. Here, τ_0 is the critical stress for β to become 1 and τ_V is the parameter representing the degree of inhomogeneity. For a homogeneous distribution, $\tau_V = 0$ and β becomes a step function of τ: when $\tau < \tau_0$, deformation does not proceed ($\beta = 0$) and when $\tau \geqq \tau_0$, unlimited deformation would occur if the distribution of obstacles remained unchanged, i.e., τ_0 were constant. Actually, however, the dislocation density increases as deformation proceeds and τ_0 increases. As a result, unlimited

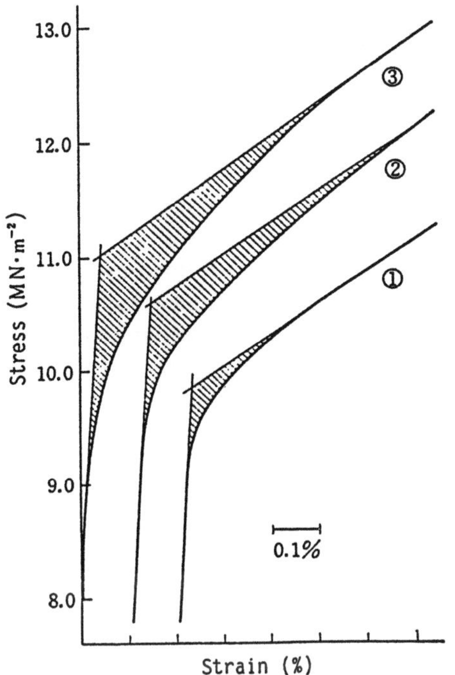

Fig. 9.10. Stress strain curves of lead obtained by a tensile test at 77 K after creeping at 293 K for various periods of time under $7\,\text{MN}\,\text{m}^{-2}$ [9.43]. Curve ①: crept for 0 s, ②: for 1920 s, ③: for 57 600 s

deformation does not occur. When the distribution is inhomogeneous, $\tau_V > 0$ and dislocations can move in the weak regions even when $\tau < \tau_0$.

As the applied stress is gradually increased, deformation proceeds with an increase in β. As τ approaches τ_0, τ_0 increases so that the relation $\tau_0 \geq \tau$ is maintained. If the work-hardening rate is θ for $\beta = 1$, then

$$h = \frac{\theta}{\beta} = \theta \exp\left(\frac{\tau_0 - \tau}{\tau_V}\right) .$$ (9.25)

Thus, if no recovery occurs, the stress-strain curve runs as shown in Fig. 9.10. The linear hardening part is due to the increase in τ_0, and the hatched area indicates the deformation of the soft region. It is seen that h becomes very large at low stresses.

At low temperatures, the soft regions are gradually consumed as deformation proceeds, and finally $\tau_V \cong 0$ and $\tau = \tau_0$. At high temperatures, however, the state $\tau_V > 0$ and the relation $\tau < \tau_0$ remains even after a large deformation, because recovery proceeds with deformation [9.41, 43–45]. In fact, Fig. 9.10, which was obtained for lead quenched to 77 K after creeping for various periods at room temperature (high temperature for lead), shows that the hatched area widens as the creep time increases. *Alden* [9.43] explained this finding from the viewpoint that a cell structure of inhomogeneous dislocation distribution is developed as the creep proceeds.

These theories are both qualitative and do not give a quantitative estimation of $M(\tau)$ or τ_V as a function of deformation conditions. However, they show that the value of h becomes very high through localization of deformation. In fact, values for h as high as μ have been measured as will be described in the next section. The temperature dependence which gave the basis of the jog-drag theory of *Hirsch* and *Warrington* (Sect. 9.1.1), can be explained by this very high h [9.41]. The internal structure in a material deformed at high temperature, where the recovery proceeds very fast, is an inhomogeneous dislocation structure, which gives a very high h. Therefore, when the material is deformed at very high temperature is deformed again at low temperature, where almost no recovery occurs, the effect of work hardening on the flow stress can be very strong even if the plastic deformation is very small. From this point of view, the experimental results obtained by *Hirsch* and *Warrington* do not necessarily mean the existence of high effective stress.

9.1.4 Experimental Values of h and r

From the Bailey-Orowan equation (8.50),

$$h = \frac{d\sigma}{d\varepsilon} + \frac{r}{\dot{\varepsilon}} \, . \tag{9.26}$$

Thus, h can be obtained by measuring the wok-hardening rate $d\sigma/d\varepsilon$ when $r = 0$ or $\dot{\varepsilon} \rightarrow \infty$.

Mitra and *McLean* [9.27] used a technique in which a specimen deformed at high temperatures is cooled to a low temperature where $r \cong 0$ and deformed further at the low temperature. In this technique, an unloading effect and a recovery effect during cooling may give an error. Further, they obtained the recovery rate r by reducing the applied stress during high-temperature creep and measuring the zero-creep period. From the relation between the stress reduction $\Delta\sigma$ and the zero-creep period Δt, r was estimated

$$r = \lim_{\Delta\sigma \rightarrow 0} \frac{\Delta\sigma}{\Delta t} \, . \tag{9.27}$$

When a sharp bend is observed on the creep curve after the stress reduction, as in Fig. 9.6, this technique is reasonable. However, in many cases, no sharp bend is observed, which may give a significant error in the experimental Δt. In spite of including this possible error, r/h obtained agrees with $\dot{\varepsilon}$ with an error of about a factor of 2.

Watanabe and *Karashima* [9.46] used a machine which permits a constant-rate tensile test and a constant-load creep test. In some cases, the cross-head of the machine was arrested during creep and r was measured as the stress relaxation rate. In other cases, during creep the test mode was changed to a tensile test with a strain rate much higher than the creep rate immediately before the change, and h was measured as the work-hardening rate immediately after the change. This technique surpasses that of *Mitra* and *McLean* in that neither the unloading effect nor the cooling effect is included, and the reported agreement of r/h with $\dot{\varepsilon}$ is very good [9.46].

Barrett et al. [9.47] threw doubt upon the very high h of 0.1–1 μ obtained by the technique and argued for the possibility that the apparent relation of $\dot{\varepsilon} \cong r/h$ is obtained in spite of the fact of incorrect h and r obtained by the above technique. However, an unrealistic assumption is included in their argument. The most serious problem to call attention to is the rigidity of the machine, though this problem is overlooked or made light of by many researchers.

Removing $\dot{\varepsilon}$ from (8.27b, 29 and 50), we obtain

$$\dot{\sigma} = \frac{K}{K+h}(h\dot{\varepsilon}_{\mathrm{a}} - r) \, . \tag{9.28}$$

Since $\dot{\varepsilon}_{\mathrm{a}} = 0$ during the stress relaxation, the relaxation rate is given by

$$\dot{\sigma}_{\mathrm{r}} = -\frac{Kr}{K+h} \, . \tag{9.29}$$

Accordingly, the equality $\dot{\sigma}_{\mathrm{r}} = -r$ does not hold unless $K \gg h$. In addition, K cannot be made larger than E. Therefore, when h is as high as E, the condition $K \gg h$ can never be realized and necessarily $|\dot{\sigma}_{\mathrm{r}}| < r$. The apparent work-hardening rate immediately after the sudden increase of strain rate is given by (8.54a). Therefore, when $\dot{\varepsilon}_{\mathrm{af}} \gg \dot{\varepsilon}_{\mathrm{a0}}$,

$$\left(\frac{d\sigma}{d\varepsilon_{\mathrm{a}}}\right)_{\mathrm{f}} = \frac{Kh}{K+h} < h \, . \tag{9.30}$$

In other words, the slope is necessarily smaller than h unless $K \gg h$. However, the ratio of these two variables,

$$\frac{-\dot{\sigma}_{\mathrm{r}}}{(d\sigma/d\varepsilon_{\mathrm{a}})_{\mathrm{f}}} = \frac{K+h}{Kh} \frac{Kr}{K+h} = \frac{r}{h} \, , \tag{9.31}$$

gives the correct ratio of r/h regardless of the value of K/h.

Conversely, if $K \ll h$, $(d\sigma/d\varepsilon_{\mathrm{a}})_{\mathrm{f}} \cong K$. Therefore, what is measured by this technique is not the aimed at h but the apparent Young's modulus K. Thus, the technique used by *Watanabe* and *Karashima* is in principle reasonable and excellent in its simplicity, but applicable only when K is at least as large as h and its value is exactly known.

If we use the strain rate change technique described in Sect. 8.3.6b, h and r can be obtained more accurately. From (9.28), the slope of the stress-strain curve immediately after the sudden change in strain rate can be expressed as

$$\left(\frac{d\sigma}{d\varepsilon_{\mathrm{a}}}\right)_{\mathrm{f}} = \frac{K}{K+h}\left(h - \frac{r}{\dot{\varepsilon}_{\mathrm{af}}}\right) \, . \tag{9.32}$$

Therefore, if the slope is measured for various $\dot{\varepsilon}_{\mathrm{af}}$'s, the linear relationship between $(d\sigma/d\varepsilon_{\mathrm{a}})_{\mathrm{f}}$ and $1/\dot{\varepsilon}_{\mathrm{af}}$ is confirmed and simultaneously h and r are deter-

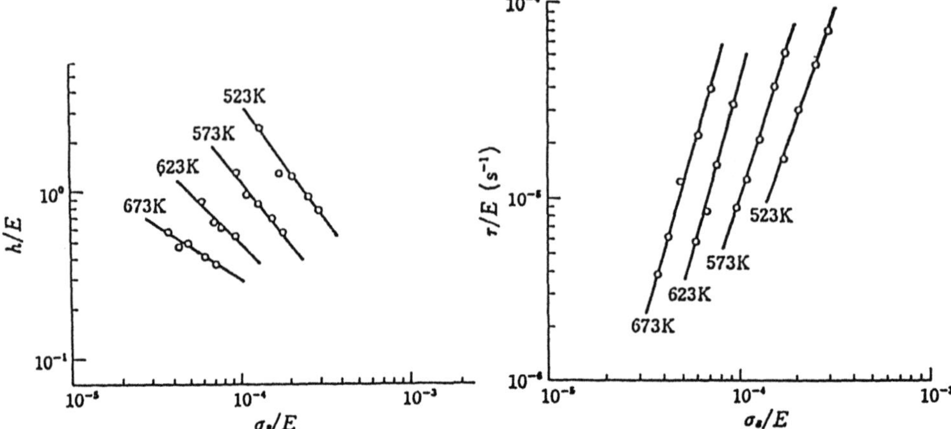

Fig. 9.11. Dependence of pure work-hardening rate h on flow stress σ_s in steady-state deformation of pure aluminum [9.48]

Fig. 9.12. Dependence of pure recovery rate r on flow stress σ_s in steady-state deformation of pure aluminum [9.48]

mined, provided again that K is at least as high as h and the value of K is exactly known.

Figures 9.11 and 9.12 show h and r obtained by *Horita* and *Yoshinaga* [9.48] using the strain-rate change technique. In their experiment, special attention was paid to increasing K to be sufficiently high and to precisely determining the value of K. Since their data points satisfy the above linear relationship, their h and r may be highly reliable.[2]

The first to use the stress change technique for the determination of h were *Ishida* and *McLean* [9.49]. In this technique, the instantaneous plastic strain $\Delta\varepsilon$ accompanying the stress change $\Delta\sigma$ is measured and the value of $\Delta\sigma/\Delta\varepsilon$ extrapolated to $\Delta\sigma \to 0$ is taken as h. *Oikawa* et al. [9.50] conducted precise experiments using this technique. According to *Nakashima* et al. [9.51], values of h and r measured by the stress change, stress relaxation and strain-rate change techniques during the high-temperature deformation of pure aluminum agree with each other within an error of 30% for h and 20% for r. All the specimens were prepared by the same process by the same batch. Their results indicate that h and r obtained by any one of the above three techniques are highly reliable.

From Figs. 9.11 and 9.12, it is seen that either h or r is proportional to a power of the stress σ; namely, h is proportional to σ^{-n_h} and r to σ^{n_r}. The values of n_h and n_r in pure aluminum obtained by *Horita* and *Yoshinaga* [9.48] are listed in Table 9.1. The exponent n_h is about 1, though it tends to decrease with rising temperature, and n_r is about 3, though it tends to increase with rising temperature.

---

[2] As shown by (8.59), h can also be determined by the stress relaxation test from the ratio of the plastic strain rates immediately before and after the start of relaxation, and r can be determined by using the relation $r = h\dot{\varepsilon}_0$. However, the reliability is inferior to that of the strain-rate change technique, because the measurement is performed for only one value of $\dot{\varepsilon}_{af} = 0$.

174

Table 9.1. Stress exponents and activation energies for $\dot{\varepsilon}$, h and r in steady-state deformation of pure aluminum [9.48]

	523 K	573 K	623 K	673 K		[kJ/mol]
n	4.0	4.4	4.8	4.3	Q	134.7
n_h	1.35	1.24	1.02	0.65	Q_h	44.6
n_r	2.96	3.12	3.84	3.70	Q_r	88.1

The exponent n_h obtained by *Mitra* and *McLean* [9.27] is 1.7 in aluminum and 1.1 in nickel, and n_r is 3.0 and 3.5, respectively, though the technique used by them is not very reliable (as described previously). The value of n_h obtained by *Sakurai* et al. [9.52] using the stress relaxation technique is 0.88 in aluminum and 1.5 in α-iron, and n_r is 4.3 and 3.2, respectively. Though n_h and n_r in lead obtained by *Oikawa* et al. [9.53] using the strain-rate change and stress relaxation techniques are very high, being 3.6 for n_h and 5.7 for n_r, their n_h obtained by the stress change technique [9.54] is 1.5. As stated above, except for some of the results obtained by *Oikawa* et al., all other results reported so far show that generally $n_h \cong 1$ and $n_r \cong 3$.

The stress exponents determined experimentally agree fairly well with the theoretical values based on the network growth model desccribed in Sect. 9.1.2. If s in (9.16) is almost independent of stress, as assumed by *H.E. Evans* and *Knowles*, the theory predicts that h is inversely proportional to stress and $n_h \cong 1$. As shown by (9.12), the theory also predicts that the recovery rate obeys the 3rd power law of stress. However, the experimental temperature dependence of h and r is hard to understand.

The experimental h and r described above were obtained in the temperature range where the activation energy evaluated from the temperature dependence of steady-state strain rate is equal to the activation energy for lattice self-diffusion. According to the network growth theory, the temperature dependence of strain rate arises from the temperature dependence of r, and h does not depend on temperature. In contradiction to this theoretical prediction, the measured h depends on temperature, and the activation energy Q_r evaluated from the temperature dependence of r is considerably lower than the activation energy for lattice self-diffusion Q_l [in Table 9.1, $Q_r \cong (2/3)Q_l$].

In pure metals, where dislocations move very fast, their moving paths are considered to be exclusively dependent on the internal structure and independent of temperature. Therefore, the work-hardening rate of internal stress, h, also should be exclusively dependent on the internal structure and independent of temperature. For example, in (9.16) when the internal structure is the same, naturally l_1 is the same, and there is no reason for the slip distance s to depend on temperature. Also in (9.22) and (9.25) based on the inhomogeneous deformation, all the factors $[M(\tau), \bar{S}, \theta, \tau_0$ and $\tau_V]$ which determine h are governed by the internal structure and stress, and do not depend on temperature. Accordingly, the observed temperature dependence of h strongly suggests that the internal structure depends on temperature even in the steady-state deformation under the same stress.

If the recovery is a single thermally activated process, a change of temperature only causes a change in the rate constant of recovery. In pure metals where the dislocation glide is an athermal process, there is no reason for the change in dislocation structure due to plastic deformation to depend on temperature, with the consequence that the internal structure in the steady state under the same stress should be independent of temperature. From these considerations, it is believed that the temperature dependence of h reflects that the recovery of the dislocation structure is not a single thermally activated process.

The areal ratio of pipe diffusion along dislocation lines to lattice diffusion is approximately $(b/l_1)^2$. This means that the contribution of pipe diffusion at the same temperature increases as l_1 decreases. According to *H.E. Evans* and *Knowles* [9.5], there is a critical link length which is given by

$$l_c = b \left[\frac{D_p}{D_l} \ln \left(\frac{\sqrt{3} l_c}{2b} \right) \right]^{1/2} , \tag{9.33}$$

where D_p and D_l are the diffusion coefficients of pipe and lattice diffusions, respectively. When $l_1 < l_c$, the contribution of pipe diffusion to the network growth prevails, while when $l_1 > l_c$, the contribution of lattice diffusion prevails.

As described in Sect. 9.1.3, when there is a link length distribution, the relative contribution of pipe and lattice diffusions depends on the local mesh size of the dislocation network. Further, the critical length l_c decreases as temperature rises. If the recovery of dislocation structure as a whole is controlled by such multiple processes, the temperature dependence of dislocation structure is understandable. *Horita* and *Yoshinaga* [9.48] showed that even in the power-law creep region of high-temperature type the observed recovery is not described by a single thermally activated process over a wide temperature range. Further, when the dislocation structure depends on temperature, r is naturally affected by the temperature dependence. Then, the activation energy obtained experimentally is only apparent and cannot be directly compared with that for a single diffusion process. This may be why the observed recovery rate does not correspond to the lattice diffusion.

When the stress is constant, as in constant-stress creep or the steady state of tensile deformation, the strain rate is expressed by the ratio of r/h as in (9.4). Then the stress exponent in (8.6), n, is given by

$$n = n_r + n_h . \tag{9.34}$$

In the same way, the activation energy for deformation, Q, is given by

$$Q = Q_r + Q_h , \tag{9.35}$$

where Q_h indicates the temperature dependence of h, because experimentally $h \propto \exp(Q_h/RT)$. As shown in Table 9.1, $n = 4.3$ and $Q = 135\,\text{kJ/mol}$ [9.55]. However, the above study indicates that the agreement of $Q \cong Q_l$ is only apparent and merely means that $(Q_r + Q_h)$ happens to approximately equal Q_l.

It seems unnatural to regard the agreement as accidental, because the equality $Q \cong Q_l$ is well known to hold widely for the high-temperature deformation of metals. Unfortunately, however, no theory has been proposed yet to show the necessity of $Q_r + Q_h \cong Q_l$.

9.2 High-Temperature Deformation Mechanism in Alloys

The strength of alloys is in most cases higher than that of that base pure metals. The strengthening depends not only on the class of alloys but also on the distribution of solute atoms. However, in many alloys the distribution becomes random at high temperature and solid solutions are formed even in alloys where the solute atoms form clusters, precipitate or arrange regularly at ambient temperature, because the contribution of the entropy term to the free energy becomes significant at high temperatures. Since a material containing other phases such as precipitates can in the widest sense be regarded as a kind of composite material, it will be described in the next chapter. In this section, the argument will be confined to solid solutions of single phase.

The principal factor of solution hardening is that the solute atoms hinder the dislocation motion. At low temperatures, solute atoms do not migrate during dislocation motion. At high temperatures, however, a region of high concentration of solute atoms will be formed around a dislocation, because solute atoms diffuse preferentially to places of low interaction energy with the dislocation. This distribution is called the solute atmosphere.

In such a case as interstitial solute atoms occupying octahedral interstitial sites in bcc metals, the interaction energy depends on the direction of the principal axes of the strain field induced by a solute atom. The distribution of the solute atoms occupying the sites such that the direction of the principal axes gives a low interaction energy is called the *Snoek atmosphere* [9.56, 57], while the concentration distribution described above is called the *Cottrell atmosphere* [9.58]. The Snoek atmosphere is of interest mainly at relatively low temperatures, because it is formed by only one jump of the solute atoms, whereas the Cottrell atmosphere is of more interest at high temperatures.

At high temperatures, asymmetric atmospheres with various concentrations may be formed around a moving dislocation, depending on the value of the drift velocity of solute atoms relative to the dislocation velocity. Then the atmosphere offers a viscous resistance to dislocation motion. This resistance acts not only against glide motion but also against climb motion of dislocations. Therefore, the atmosphere may affect the multiplication and annihilation rates of dislocations and affect the density and arrangement of dislocations in steady-state deformation. First, the deformation behavior particular to alloys will be described.

9.2.1 High-Temperature Deformation Behavior of Solution-Hardened Alloys

The flow stress of solution-hardened alloys generally shows a temperature dependence as shown in Fig. 9.13. The data in the figure were obtained by *Asada* et al. [9.59], showing how the critical resolved shear stress in Al-Mg alloy single crystals depends on temperature. A notable feature of this kind of alloy is that the flow stress increases with rising temperature in the intermediate temperature region B. Since the temperature dependence is the inverse of the usually observed ones, it will be called inverse temperature dependence.

In the intermediate temperature region, serrations are frequently observed in the stress-strain curve. Figure 9.14 is an example. This phenomenon is sometimes

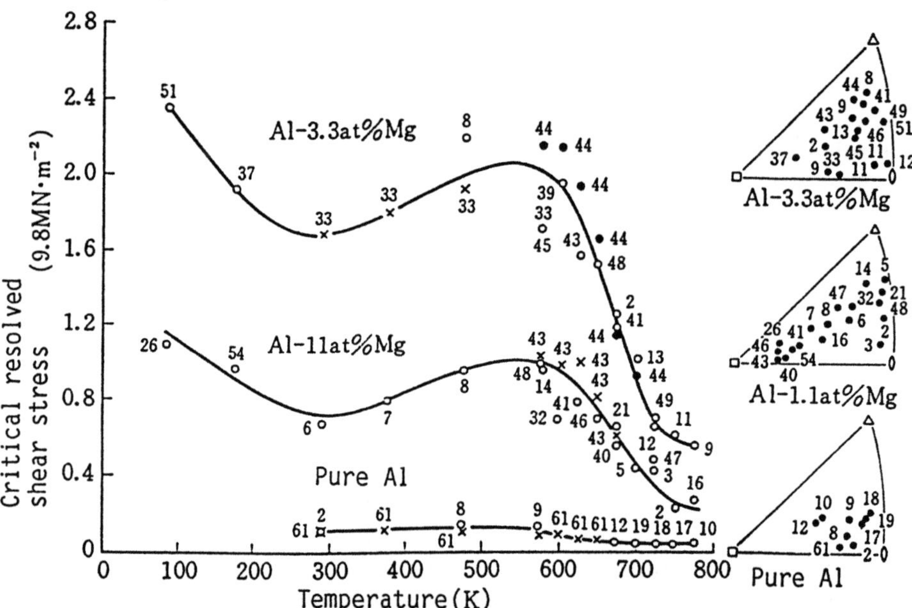

Fig. 9.13. Temperature dependence of critical resolved shear stress in pure aluminum and Al-Mg alloy single crystals deformed by tension at $4.17 \times 10^{-4}\,\mathrm{s}^{-1}$ [9.59]

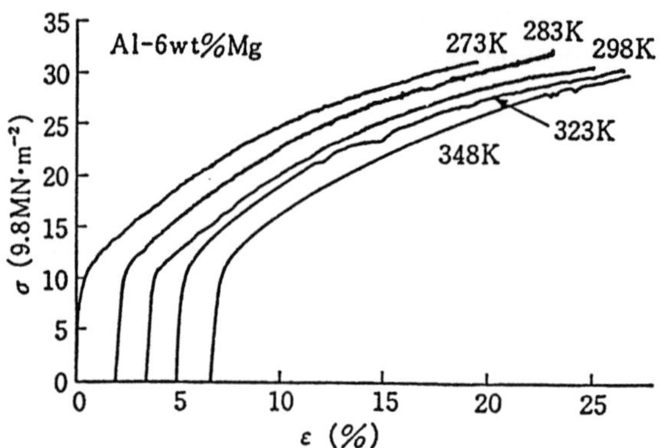

Fig. 9.14. PL effect observed in an Al-Mg alloy [9.60]

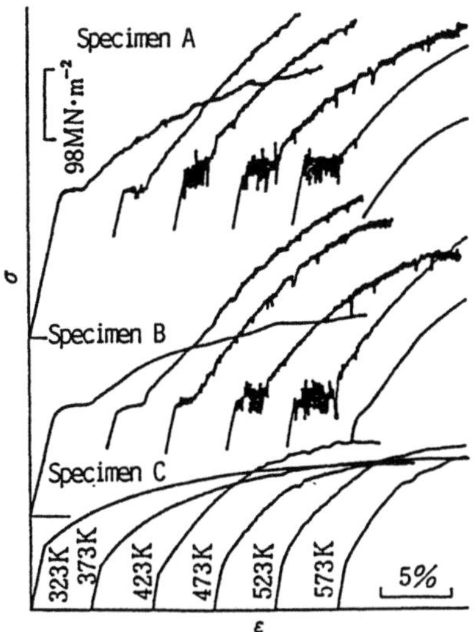

Fig. 9.15. PL effect observed in αFe-C [9.61]. Specimen A: C 250 ppm, N 14 ppm Specimen B: C 50 ppm, N 13 ppm Specimen C: C+N $\leqq$ 10 ppm

referred to as dynamic strain aging, because it is considered to arise from strain aging occurring in a dynamical state of plastic deformation. However, it is called here the Portevin-LeChatelier effect (PL effect), after the discoverers of this phenomenon.

Solute atoms occupying the octahedral interstitial sites in bcc metals, such as carbon in α-iron, have an especially high ability for solution hardening, and bring about a marked PL effect. Figure 9.15 shows an example. In interstitial solid solutions, when the PL effect occurs, the work hardening becomes especially large, as seen in this example, whereas in substitutional solid solutions the work hardening accompanying the PL effect is not so remarkable, as shown in Fig. 9.14.

Further, in the intermediate temperature region, when the strain rate is lowered, the work-hardening rate increases and thus the flow stress increases. Figure 9.16 shows an example. This strain rate dependence of flow stress is the inverse of usually observed ones, and corresponds to the inverse temperature dependence. These inverse dependences suggest that the obstacles to dislocation motion are formed by a thermally activated process.

Figure 9.17 shows stress-strain curves obtained by tensile deformation of various aluminum alloys in the high temperature region C in Fig. 8.9. It is seen that a work softening occurs in highly solution-hardened alloys. The softening characteristic of the high-temperature deformation of these alloys was found by *Horiuchi* et al. [9.63] and called the high temperature yield point phenomenon.

In contrast to the dislocation structure in pure metals deformed at high temperatures where the dislocations distribute very inhomogeneously as shown in Fig. 9.7, the structure in highly solution-hardened Al-Mg alloys is relatively homogeneous, as seen in Fig. 9.18. Further, the stress exponent n in (8.6) decreases from about 5, typical for pure metals, to 3, typical for solution-hardened alloys, as the solute concentration increases (Fig. 9.19). However, the activation energy

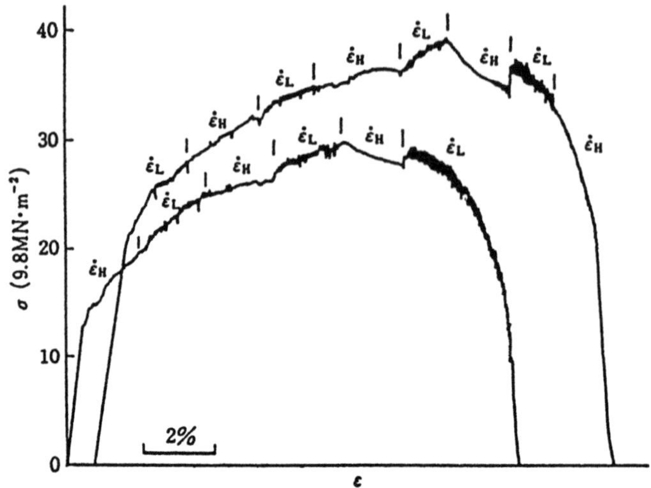

Fig. 9.16. Effect of strain-rate change on stress-strain curve of vanadium containing 170 ppm carbon and 42 ppm oxygen as impurities [9.62]. Test temperature: 573 K; higher strain rate, $\dot{\varepsilon}_H$: 1.3 × $10^{-3}\,\mathrm{s}^{-1}$; lower strain rate, $\dot{\varepsilon}_L$: 1.3 × $10^{-4}\,\mathrm{s}^{-1}$

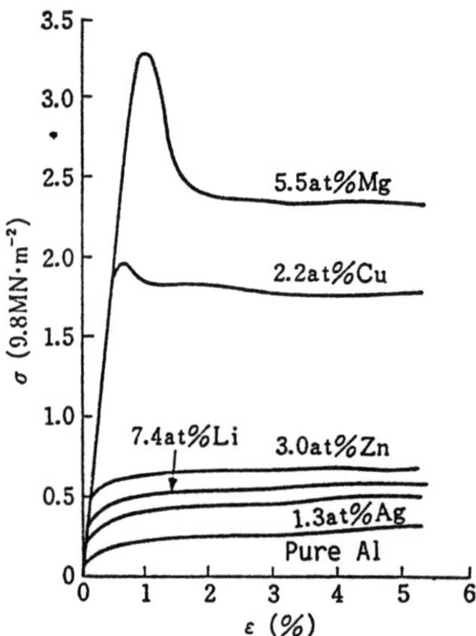

Fig. 9.17. Stress-strain curves of various aluminum alloys deformed by tension at 773 K and 1.3 × $10^{-2}\,\mathrm{s}^{-1}$ [9.63]

for high-temperature deformation is approximately the same as that for lattice self-diffusion, as is the case in pure metals. *Sherby* and *Burke* [9.66] classified solid-solution alloys according to their stress exponent, calling cases where $n \cong 3$ class I alloys and cases where $n \cong 5$ class II alloys. Solution hardening is high in class I alloys but low in class II alloys. The deformation behavior of class II alloys is similar to that of pure metals.

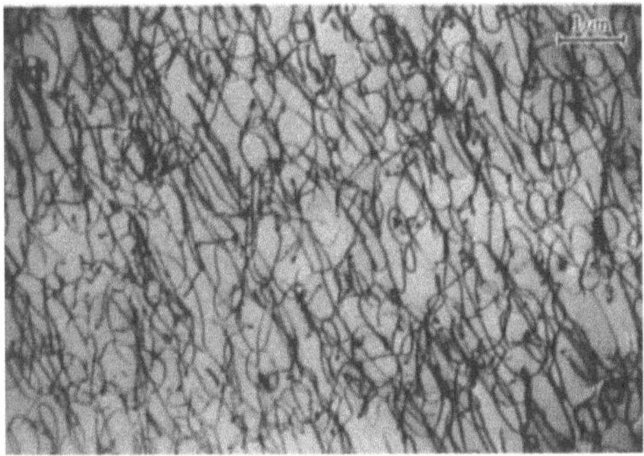

Fig. 9.18. Dislocation structure in an Al-5.1 at.% Mg alloy frozen from steady-state creep at 632 K under 47 MN m^{-2} [9.64]

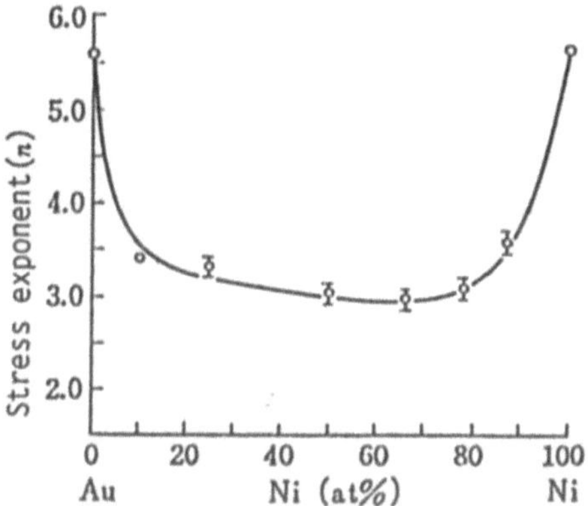

Fig. 9.19. Concentration dependence on stress exponent n in steady-state creep of Au-Ni alloys, which shows that n decreases from about 5 to 3 by alloying [9.65]

In the following, it will be shown that the above deformation behavior characteristic of alloys can be explained by considering the interaction between solute atoms and dislocations.

9.2.2 Drift Flow of Solute Atoms Relative to a Moving Dislocation

Various sources can be considered for the interaction between solute atoms and dislocations [9.67]. Of them, we consider here the size effect (see Sect. 3.4) because this effect is especially large in the solution hardening. The characteristics of the interaction at high temperatures will be described in terms of this effect.

The interaction energy of a solute atom with an edge dislocation is given by (3.23). The interaction with a screw dislocation is much weaker than that with an edge dislocation in substitutional solid solutions [9.68]. On the other hand, a solute atom in an octahedral interstitial site in bcc metals interacts strongly also with a screw dislocation [9.56, 57]. However, the interaction has the same character as that in substitutional solid solutions in that the interaction energy W decreases inversely proportional to the distance from the dislocation.

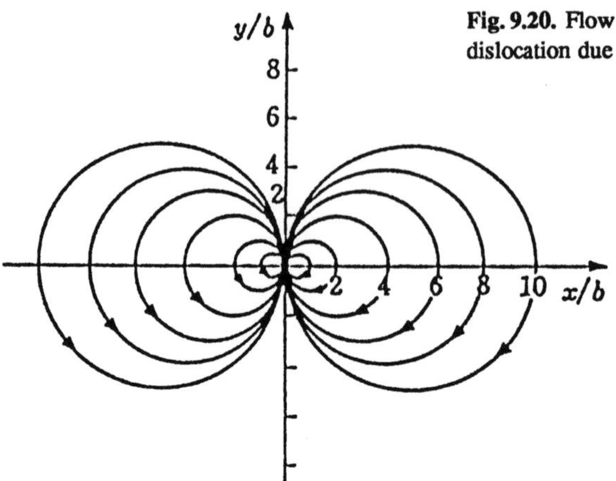

Fig. 9.20. Flow of solute atoms around a stationary dislocation due to the interaction given by (3.23)

The flow of solute atoms caused by the interaction is given by Einstein's equation

$$J_W = -\frac{D_c}{k_B T} c \nabla W \,,$$ (9.36)

where D_c is the diffusion coefficient of solute atoms, c the concentration, and ∇W the gradient of the interaction energy.

When the dislocation is stationary, the flow of solute atoms due to the interaction given by (3.23) draws a circular trajectory as shown in Fig. 9.20 [9.69]. When the dislocation is moving with a velocity v, an apparent flow relative to the dislocation

$$J_v = -cv$$ (9.37)

is added to the above flow. In short, the flow of solute atoms relative to the moving dislocation may be expressed as

$$J_d = J_W + J_v = -\frac{D_c}{k_B T} c \nabla \left(W + \frac{k_B T v}{D_c} x \right) \,.$$ (9.38)

This flow is the same as that relative to the stationary dislocation, provided that the interaction energy is now changed to an apparent one,

$$W_a = W + \frac{k_B T v}{D_c} x \,.$$ (9.39)

The flow line intersects the equi-W_a curves perpendicularly. Therefore, when W is given by (3.23), the equation representing the flow line becomes

$$A \frac{x}{x^2 + y^2} + \frac{k_B T v}{D_c} y = \xi \,,$$ (9.40)

which gives flow lines as shown in Fig. 9.21 [9.70], where ξ is a parameter. In the

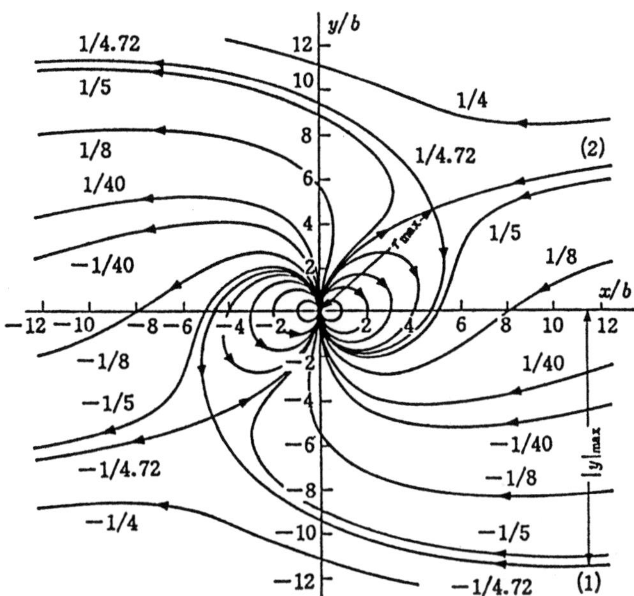

Fig. 9.21. Flow of solute atoms relative to a moving dislocation with a velocity $v = 2.24 \times 10^{-2} A D_c / k_\mathrm{B} T b^2$. The numbers indicate the values of the parameter ξ in units of A/b [9.70]

above, the direction of the edge dislocation line is taken parallel to the z-axis, the direction of the glide motion parallel to the x-axis, and the direction normal to the slip plane parallel to the y-axis. The origin is fixed at the moving dislocation, the position of solute atom is (x, y), and $3\mu b \Omega \varepsilon_a / \pi$ in (3.23) is replaced by A.

If the equation of the flow line is expressed in polar coordinates and solved with respect to r we obtain

$$r = \frac{\xi \pm \sqrt{\xi^2 - (2A k_\mathrm{B} T v / D_c)\, \sin 2\theta}}{(2 k_\mathrm{B} T v / D_c)\, \sin \theta} \ . \tag{9.41}$$

When

$$|\xi| \gtreqqless \xi_c = \left(\frac{2A k_\mathrm{B} T v}{D_c} \right)^{1/2} , \tag{9.42}$$

the double signs $\pm$ in (9.41) correspond to open flow lines, which do not go through the dislocation core, and closed flow lines, both ends of which are at the core, respectively. When $|\xi| < \xi_c$, the flow lines have one end at the core.

The boundaries separating the two regions of open flow lines and closed or one-end flow lines are expressed as the two open flow lines of $|\xi| = \xi_c$ and represented by

$$y = \pm \frac{A[1 + (1 - \sin 2\theta)^{1/2}]}{\xi_c} \ . \tag{9.43}$$

Accordingly, the region of solute atoms flowing into the dislocation core is

183

limited by

$$-\frac{(1+\sqrt{2})A}{\xi_c} < y < \frac{2A}{\xi_c} .$$

In the same way, the region where solute atoms flow out of the core is given by

$$-\frac{2A}{\xi_c} < y < \frac{(1+\sqrt{2})A}{\xi_c} .$$

Therefore, there is a critical distance given by

$$|y|_{\max} = \left(1 + \frac{1}{\sqrt{2}}\right)\left(\frac{AD_c}{k_BTv}\right)^{1/2} , \tag{9.44}$$

and the solute atoms more than $|y|_{\max}$ away from the slip plane always flow apart from the dislocation core.

When the solute concentration is uniform everywhere, the change in concentration at an arbitrary locus is given by

$$\frac{\partial c}{\partial t} = -\nabla \boldsymbol{J}_d = -\nabla \boldsymbol{J}_w = \frac{D_c}{k_BT}c\nabla^2 W . \tag{9.45}$$

However, $\nabla^2 W = 0$ for the interaction energy of (3.23), and thus no change in concentration is induced by the flow. The reason for a solute atmosphere being formed around a dislocation is that there are flow lines which have their ends at the dislocation core. Accordingly, no solute atmosphere will be formed around a dislocation moving so fast that $|y|_{\max} < b$. In other words, around a dislocation moving faster than the critical velocity which is given from (9.44) by

$$v_{\mathrm{CL}} = \frac{3+2\sqrt{2}}{2}\frac{AD_c}{k_BTb^2} \tag{9.46}$$

no solute atmosphere will be formed [9.70]. This critical condition holds as long as the interaction energy is inversely proportional to the distance from the dislocation, except that the coefficient on the right-hand side of (9.46) depends on the form of the interaction energy [9.71].

9.2.3 Resistance to Dislocation Motion due to Solute Atmosphere

When a dislocation is moving with a velocity $v < v_{\mathrm{CL}}$, a solute atmosphere will be formed around it, and then a concentration gradient is produced. Accordingly, a flow component due to the gradient,

$$\boldsymbol{J}_c = -D_c\nabla c \tag{9.47}$$

is added to $\boldsymbol{J}_d$. As a result, the flow of solute atoms is now given by

$$\boldsymbol{J} = \boldsymbol{J}_c + \boldsymbol{J}_W + \boldsymbol{J}_v = -\left(D_c\nabla c + \frac{D_c}{k_BT}c\nabla W + cv\right) . \tag{9.48}$$

In a steady state, $\nabla J = 0$, namely

$$\nabla \left(D_c \nabla c + \frac{D_c}{k_{\mathrm{B}} T} c \nabla W + c v \right) = 0 . \tag{9.49}$$

If this equation can be solved, the concentration distribution around the dislocation is known, and thus the resistance the dislocation is offered by the solute atmosphere can be calculated.

When the atomic volume is Ω and the solute concentration c is expressed in an atomic ratio, the number of solute atoms per unit volume is c/Ω. A solute atom is attracted by a dislocation with a force $f = -\nabla W$, and inversely, the dislocation is attracted by a solute atom with a force $-f$. Accordingly, a force f per solute atom is necessary to move the dislocation forward. The force from all the solute atoms is given by the volume integral

$$F = - \int \frac{c}{\Omega} \nabla W dV . \tag{9.50}$$

For a straight dislocation, the solute concentration will not change in the direction parallel to the dislocation. The force per unit dislocation length in the direction of v is given by the area integral

$$\tau_{\mathrm{d}} b = - \int \frac{c \nabla W}{\Omega} dS \cdot \left(\frac{v}{v} \right) , \tag{9.51}$$

where τ_{d} is the shear stress required to move the dislocation with a velocity v and is called the drag stress of the solute atmosphere.

It is difficult to obtain the analytical solution of the concentration distribution. *Cottrell* and *Jaswon* [9.58] gave the solution of (9.49) as the sum of an infinite series of Mathew functions with respect to the dislocation velocity. To obtain concrete values, however, some approximation is inevitable for the calculation. Further, they showed that the concentration for small v is given by

$$c = c_0 [1 + O(v)] \exp \left(-\frac{W}{k_{\mathrm{B}} T} \right) \tag{9.52}$$

where c_0 is the concentration far from the dislocation, $O(v)$ is a term first order in the infinitesimal of v and dependent on the position relative to the dislocation, being zero far from the dislocation. In a range of very small v, τ_{d} was given by *Cottrell* and *Jaswon* as

$$\tau_{\mathrm{d}} = B v , \tag{9.53}$$

where

$$B \cong \frac{17}{4} \frac{A^2 c_0}{D_c k_{\mathrm{B}} T b \Omega} \tag{9.54}$$

for the interaction energy of (3.23) [9.58].

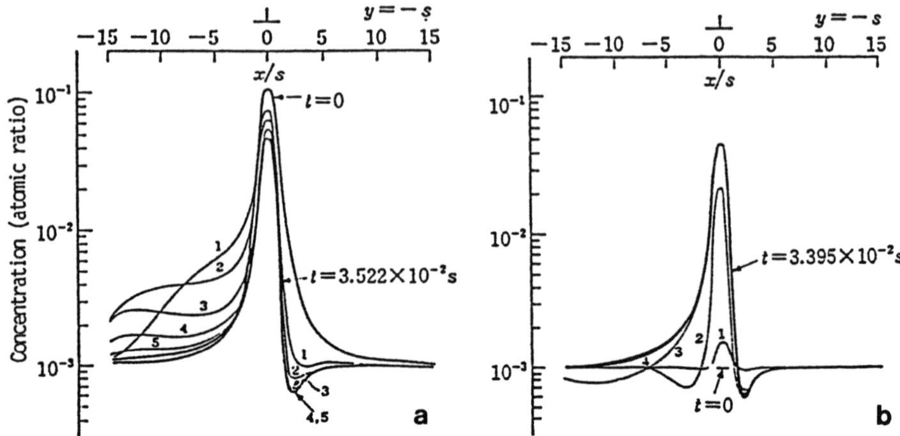

Fig. 9.22a,b. An example of change in concentration distribution accompanying dislocation movement, calculated for $A = 1.40 \times 10^{-27}\,\mathrm{J\,m}$, $c_0 = 10^{-3}$, $D_c = 1.74 \times 10^{-17}\,\mathrm{m^2\,s^{-1}}$ and $v = 2 \times 10^3\,\mathrm{\AA\,s^{-1}}$ [9.72]. Bold lines indicate the initial and steady-state distributions. (a) A saturated atmosphere around a stationary dislocation is given as the initial distribution, and the curves 1–5 are the distributions after $t = 3.807 \times 10^{-3}$, 7.894×10^{-3}, 1.434×10^{-2}, 2.031×10^{-2}, and 2.667×10^{-2} s, respectively. (b) No atmosphrere is given as the initial distribution, and the curves 1–4 are the distributions after $t = 1.479 \times 10^{-5}$, 1.748×10^{-3}, 1.163×10^{-2}, and 2.077×10^{-2} s, respectively

Yoshinaga and *Morozumi* [9.72] calculated the change in solute concentration around a moving dislocation by pursuing the jumping process of solute atoms between lattice points with a computer simulation. Two cases were studied as the initial condition: in one a uniform solute concentration was given, and in the other an atmosphere in steady state around a stationary dislocation was given. In the former case, an atmosphere is formed around a dislocation moving with a constant velocity, while in the latter case the steady-state atmosphere collapses as the dislocation proceeds. Finally the atmosphere in the two cases converge to the same steady-state one for the moving dislocation. Figure 9.22 shows the changing process as time elapses. For convenience, only the concentration distribution on the atomic plane just below the dislocation is shown in the figure.

Figure 9.23 shows the steady-state concentration distribution as a function of dislocation velocity. In this figure also, only the distribution on the plane just below the dislocation is shown. It is seen that the peak concentration decreases as the dislocation velocity increases and a region of high concentration is formed like the tail of a comet round a fast moving dislocation. It is also shown that the concentration just ahead of the dislocation decreases below the average when the dislocation velocity becomes high, because the solute atoms just ahead of the dislocation are attracted by the dislocation, while the supply of solute atoms to this locus from the outside region is insufficient to compensate the decrease.

Figure 9.24 shows a steady-state atmosphere in a KCl-KBr solid solution calculated by *Sakamoto* [9.73], where all the features of the atmosphere are shown.

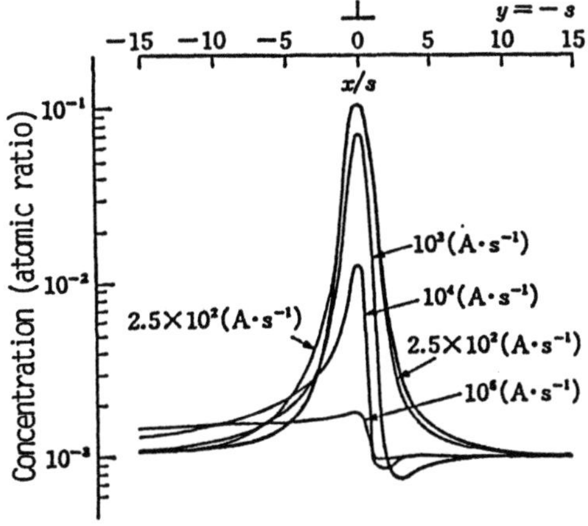

Fig. 9.23. Dislocation-velocity dependence of steady-state concentration distribution [9.72]. The parameters used here are the same as those in Fig. 9.22

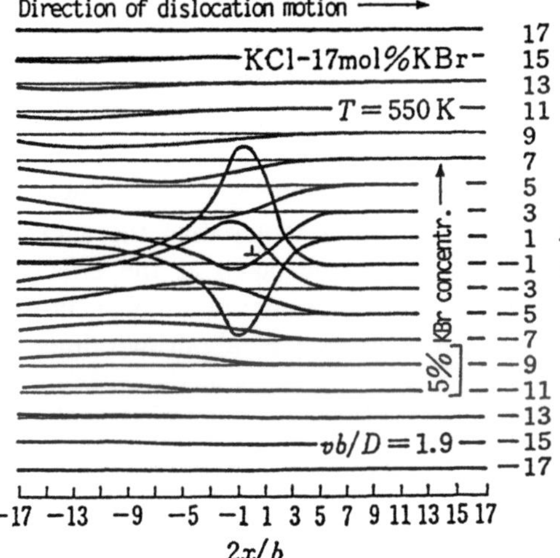

Fig. 9.24. All the features of steady-state concentration distribution around a moving dislocation, calculated by *Sakamoto* [9.73]

The drag resistance calculated from the above concentration distributions at first increases, passes a peak and then decreases as the dislocation velocity increases, as shown in Fig. 9.25. Until the velocity reaches v_{CH}, which corresponds to the maximum resistance, the center of gravity of the atmosphere deviates backward, but the size of the atmosphere does not differ much from that around a stationary dislocation. The τ_d in a region of $v \ll v_{CH}$ is almost linear with v as *Cottrell* and *Jaswon*'s equation (9.53) predicts.

In the above calculations, the range of integral (9.51) is artificially limited. *Hirth* and *Lothe* [9.74] pointed out, using the analogy of the solution to a one-dimensional approximation, that the effective range of integral probably diverges infinitely as v approaches zero. According to their analysis, the coefficient B in (9.53) becomes approximately

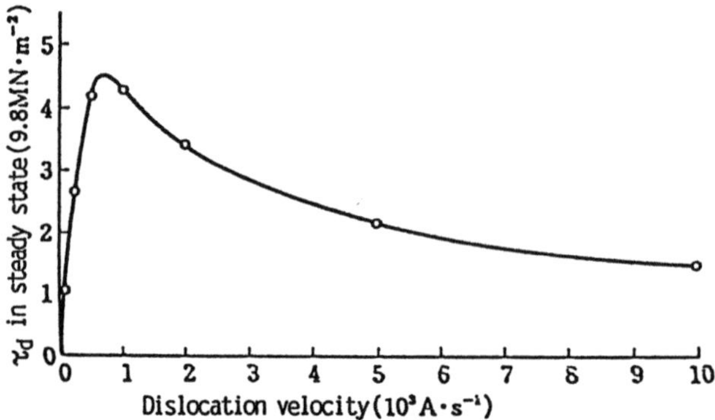

Fig. 9.25. Dislaocation-velocity dependence of solute-atmosphere drag stress in the steady state [9.72]. The parameters are the same as those in Fig. 9.22

$$B' \cong \frac{A^2 c_0}{D_c k_{\mathrm{B}} T b \Omega} \ln \frac{r_2}{r_1} , \tag{9.55}$$

where

$$r_1 = \frac{A}{k_{\mathrm{B}} T} , \qquad r_2 = \frac{D_c}{v} . \tag{9.56}$$

Form this, the effective range of integral, r_2, becomes infinite as $v \to 0$. However, when the dislocation motion is slow and r_2 becomes larger than the average spacing of dislocations, the average spacing should be taken as the effective range of integration. Then B' becomes a constant in a range of small v. In this case, as $r_2/r_1 \cong 10^3$–10^4, $\ln(r_2/r_1) = 7$–9, which gives a B' about two times as large as B.

9.2.4 Interpretation of High-Temperature Deformation Behavior of Alloys

It will be shown in the following how the deformation behavior characteristic of alloys described in Sect. 9.2.1 can be understood from the above fundamental theory.

a) PL Effect and Inverse Dependence of Flow Stress on Temperature and Strain Rate. Either of the two critical velocities, v_{CL} and v_{CH}, introduced in the preceding section increases in proportion to the diffusing velocity of solute atoms. Accordingly, the dependence of drag stress on dislocation velocity changes with temperature as shown in Fig. 9.26.

No atmosphere is formed for $v > v_{\mathrm{CL}}$. Even in this case, however, a moving dislocation is resisted by the randomly distributed solute atoms. This low-temperature resistance increases with increasing dislocation velocity [9.73] and is shown in the figure as dotted curves. Now, let us consider the resistance to a dislocation moving with a velocity v_1. In the temperature range, $T < T_2$, in the figure, the resistance decreases as temperature rises, and increases as the

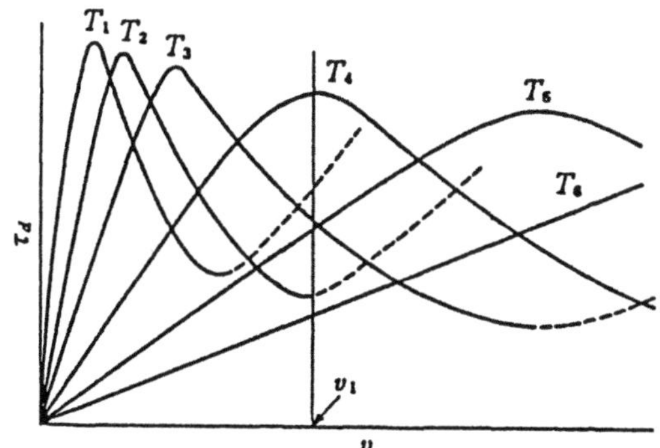

Fig. 9.26. Schematic representation of temperature dependence of the relation between dislocation velocity and drag stress [9.75]. $T_i < T_{i+1}$

dislocation velocity increases (normal temperature and strain-rate dependences). This temperature range corresponds to the low-temperature region A in Fig. 8.9. In the temperature range, $T_2 < T < T_4$, on the other hand, the resistance (drag stress) increases with rising temperature and decreases with increasing dislocation velocity (inverse temperature and strain-rate dependences). This is because the solute atmosphere resisting the dislocation motion is formed by a thermally activated process (diffusion). This temperature range corresponds to the intermediate temperature region B in Fig. 8.9. In this region, the PL effect may be observed, because $d\tau_d/dv < 0$ and thus dislocations are accelerated, resulting in a plastic instability. In the high-temperature region C of $T > T_4$, the deformation recovers stability, and for $v_1 \ll v_{CH}$ the drag stress becomes proportional to the dislocation velocity.

The above argument is based on the assumption of the uniformity of dislocation velocities. Actually, however, the assumption may not hold. According to the local fluctuation of internal stress in crystals, dislocation velocities are not thought to be uniform. Thus, even when the average velocity of dislocations is lower than v_{CH}, there can be dislocations moving faster than v_{CH}. Then the alloy may exhibit the deformation behavior typical in the intermediate temperature region. Around slower dislocations large atmospheres will be formed and the dislocations are decelerated further. Thus immobilized dislocations are accumulated, causing the high work hardening in the intermediate temperature region. This work hardening is especially high in interstitial solid solutions such as αFe-C, but low in substitutional solid solutions such as Al-Mg. There are two possibilities to be considered as the reason for this difference. One is that the screw dislocations in the former are also strongly locked by solute atmospheres, while in the latter the locking for screw dislocations is much weaker than that for edge dislocations. The other is that in the former case the recovery effect due to dislocation climb is slight in the intermediate temperature region because $D_c \gg D_1$, whereas in the latter case the recovery effect is strong even in the intermediate temperature region because $D_c \cong D_1$, resulting in a small dislocation accumulation.

b) High-Temperature Yield Point Phenomenon. In well-annealed solid solutions, it is very unlikely that the obstacles to dislocation motion decrease by deformation. Therefore, it may be reasonable to consider that a kind of work softening, called the high-temperature yield point phenomenon, arises from an increase in deformation carriers. The experimental results given in Fig. 9.27 show well the correctness of this consideration. Curve ① in the figure is the stress-strain curve obtained at a strain-rate of $\dot{\varepsilon}_a = 8.67 \times 10^{-3}\,\mathrm{s}^{-1}$. At the end of the curve, the specimen was unloaded and deformed again at a reduced strain rate of $4.17 \times 10^{-4}\,\mathrm{s}^{-1}$, which is 1/20 of the former strain rate. Curve ② is the stress-strain curve thus obtained. Curve ③ is that obtained by increasing the strain rate to the original one at the end of curve ②. The curve obtained for a virgin specimen at the lower strain rate is curve ①′. The dislocation density at an early stage of ② is considered to be nearly the same as the steady-state density for the higher strain rate, and must be much higher than the initial density of ①′. Nevertheless the flow stress is lower than that of ①′. This means that, contrary to pure metals and less solution-hardened alloys, the flow stress decreases as the dislocation density increases. Curve ② shows that the flow stress increases as the dislocation density decreases from the steady-state value for the higher strain rate to that for the lower strain rate. Conversely, curves ① and ①′ show that softening occurs because the dislocation density is increased by deformation. In the steady state the flow stresses of ①′ and ② agree with each other, showing that the dislocation density is definitely determined in the steady state, regardless of the previous history of deformation. The observation that the peak of curve ③ is lower than that of curve ① is understood by considering that the initial dislocation density of ③ has been increased by the previous deformation of ① and ② over that in the annealed state.

Figure 9.28 shows that the upper yield point recovers gradually as the dislocation density is decreased by annealing after having introduced a high dislocation

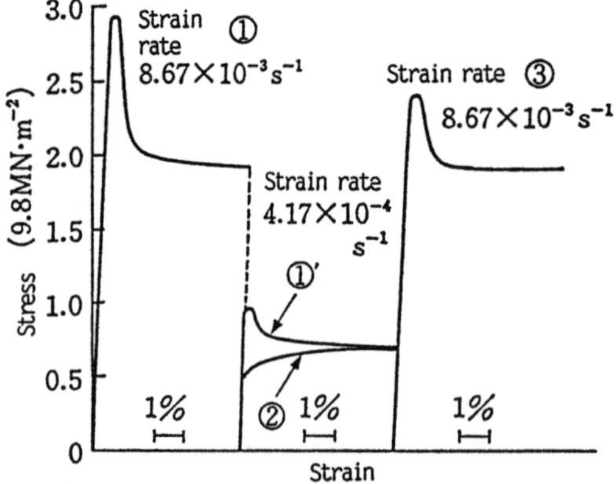

Fig. 9.27. Effect of initial dislocation density on stress-strain curve of Al-5.5at% Mg alloy deformed at 773 K [9.68]. Curves ①, ② and ③ were obtained using a single specimen and tensile tests were conducted in this order. Curve ①′ was obtained at the lower strain rate, using a virgin specimen

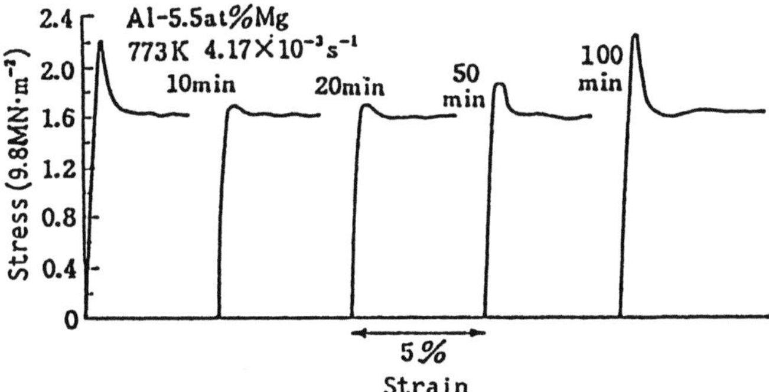

Fig. 9.28. Recovery of high-temperature yield point phenomenon by annealing [9.63]. Annealing temperature is the same as the deformation temperature, 773 K

density by the deformation to the steady state. Now let us consider the relation between dislocation density and flow stress.

When the viscous resistance of the solute atmosphere to dislocation motion is high, most of the dislocations are considered to be mobile. In other words, the mobile dislocation density ϱ_m is nearly equal to the total dislocation density ϱ.

In substitutional solid solutions, the interaction of a solute atom with a screw dislocation is weak. Thus the screw dislocations will move much faster than the edge dislocations. Assuming that the dislocation loops consist of pure edge and pure screw parts, we obtain for the strain rate

$$\dot{\gamma} = (\varrho_e \bar{v}_e + \varrho_s \bar{v}_s)b = \varrho_e \bar{v}_e b \left(1 + \frac{\varrho_s \bar{v}_s}{\varrho_e \bar{v}_e}\right) , \tag{9.57}$$

where suffix e denotes edge dislocation and suffix s, screw dislocation.

The length of edge dislocation L_e is proportional to the velocity of screw dislocations $\bar{v}_s$, and the length of screw dislocations L_s is proportional to the velocity of edge dislocations $\bar{v}_e$ [9.76]. Accordingly

$$\frac{L_e}{L_s} = \frac{\varrho_e}{\varrho_s} = \frac{\bar{v}_s}{\bar{v}_e} \tag{9.58}$$

or

$$\varrho_s \bar{v}_s = \varrho_e \bar{v}_e . \tag{9.59}$$

From $\bar{v}_s \gg \bar{v}_e$ and (9.58), $\varrho_e \gg \varrho_s$, i.e., $\varrho_e \cong \varrho$. Thus (9.57) can be approximated by

$$\dot{\gamma} = 2\varrho b \bar{v}_e . \tag{9.60}$$

Since the proportional relationship of (9.53) holds between τ_d and v_e,

$$\dot{\gamma} = \frac{2\varrho b \bar{\tau}_d}{B} \tag{9.61a}$$

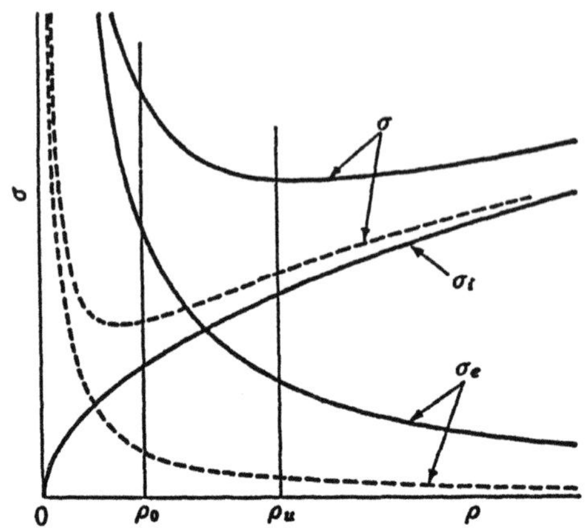

Fig. 9.29. Schematic representation of the dependence of internal stress σ_i, effective stress σ_e and flow stress σ on dislocation density ϱ [9.77]. Solid curves indicate a case of high $\dot{\varepsilon}/B$, and broken curves, a case of low $\dot{\varepsilon}/B$

or, denoting the orientation factor by $\phi(\cong 1/2 - 1/3)$,

$$\dot{\varepsilon} = \frac{2\phi^2 \varrho b \bar{\sigma}_\mathrm{d}}{B} .$$ (9.61b)

This drag stress $\bar{\sigma}_\mathrm{d}$ is experimentally measured as the effective stress $\bar{\sigma}_\mathrm{e}$.

If the internal stress is given by (8.14), by using (9.61a) the flow stress is expressed as

$$\tau = \frac{B\dot{\gamma}/2b}{\varrho} + \alpha\mu b\sqrt{\varrho} .$$ (9.62)

The effective stress (the first term on the right-hand side) decreases and the internal stress (the second term) increases with an increase in ϱ. From the condition $d\tau/d\varrho = 0$, the flow stress becomes a minimum at a critical value of ϱ, which is given by

$$\varrho_\mathrm{c} = \left(\frac{B\dot{\gamma}}{\alpha\mu b^2}\right)^{2/3} .$$ (9.63)

Figure 9.29 schematically shows this relationship between dislocation density and flow stress.

Denoting the initial dislocation density by ϱ_0 and the steady-state density by ϱ_u, work-hardening occurs for $\varrho_0 > \varrho_\mathrm{c}$, at first work softening and then work hardening for $\varrho_0 < \varrho_\mathrm{c} < \varrho_u$, and only work softening for $\varrho_u < \varrho_\mathrm{c}$. The critical density ϱ_c is proportional to $(B\dot{\gamma})^{2/3}$ and from (9.54) and (9.55) B is proportional to $A^2 c_0/D_\mathrm{c}$. Therefore, when the parameter A (which is a measure of the interaction of a solute atom with a dislocation) increases, when the solute concentration is higher, or when $\dot{\gamma}$ increases, ϱ_c becomes larger. Thus, it is expected that a conspicious work softening will be observed in an alloy under conditions where the solution hardening is high. This assumption is supported by the experimental results shown in Fig. 9.17 and 9.32.

From (9.62)

$$\dot{\gamma} = \frac{2b\varrho}{B}(\tau - \alpha\mu b\sqrt{\varrho}) \, . \tag{9.64}$$

Therefore, from the condition $d\dot{\gamma}/d\varrho = 0$, the critical dislocation density giving the maximum strain rate in a constant stress creep test is expressed by

$$\varrho_c' = \left(\frac{2\tau}{3\alpha\mu b}\right)^2 \, . \tag{9.65}$$

when $\varrho < \varrho_c'$ or

$$\tau > \frac{3}{2}\bar{\tau}_i \quad \text{or} \quad \bar{\tau}_e > \frac{1}{2}\bar{\tau}_i \, , \tag{9.66}$$

$d\dot{\gamma}/d\varrho > 0$, which is the inverse transient creep corresponding to the work softening in a tensile test. The nomenclature "inverse transient creep" comes from the behavior that the creep rate increases as deformation proceeds, as shown in Fig. 9.30, which is the inverse of the behavior normally observed in pure metals and class II alloys, where the creep rate decreases as deformation proceeds in the transient stage.

The transient creep is of a normal type when $\varrho_0 > \varrho_c'$, S-letter type [9.78] when $\varrho_0 < \varrho_c' < \varrho_u$, and inverse type when $\varrho_u < \varrho_c'$. As is known from (9.65), ϱ_c' increases in proportion to τ^2. Therefore, it is expected that the type of transient creep changes in the above order as the creep stress is increased. Against the expectation, however, Fig. 9.31 shows that a clear inverse transient is observed under a low stress and changes to S-letter type and then to the normal one as stress is increased. Though the reason for this disagreement is not clear, it may

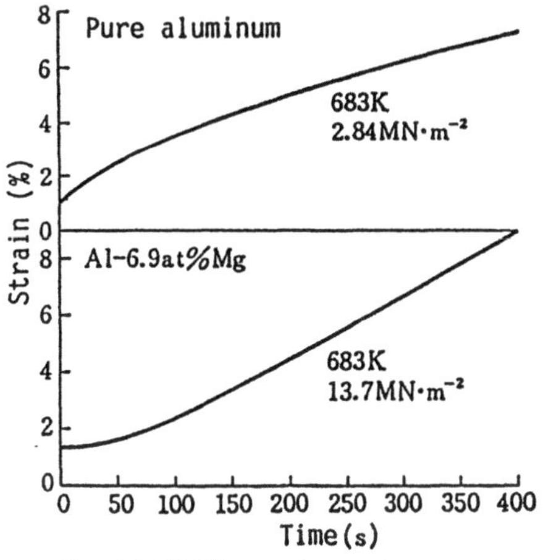

Fig. 9.30. Difference in transient creep between pure aluminum and a solution-hardened Al-Mg alloy [9.64]

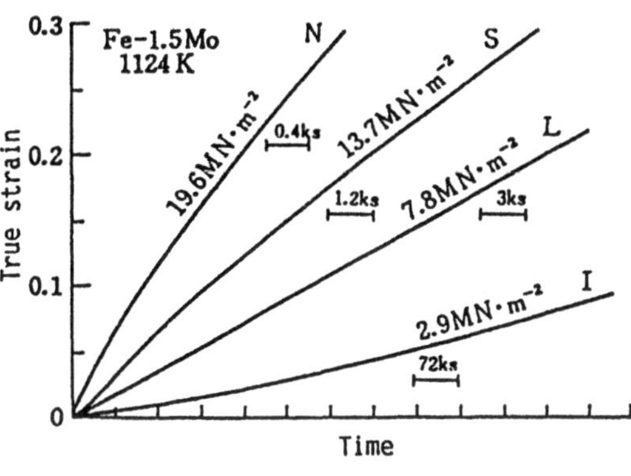

Fig. 9.31. Stress dependence of transient creep in an Fe-1.5 at.% Mo alloy [9.78], I: inverse type, L: linear type, S: S-letter type, and N: normal type

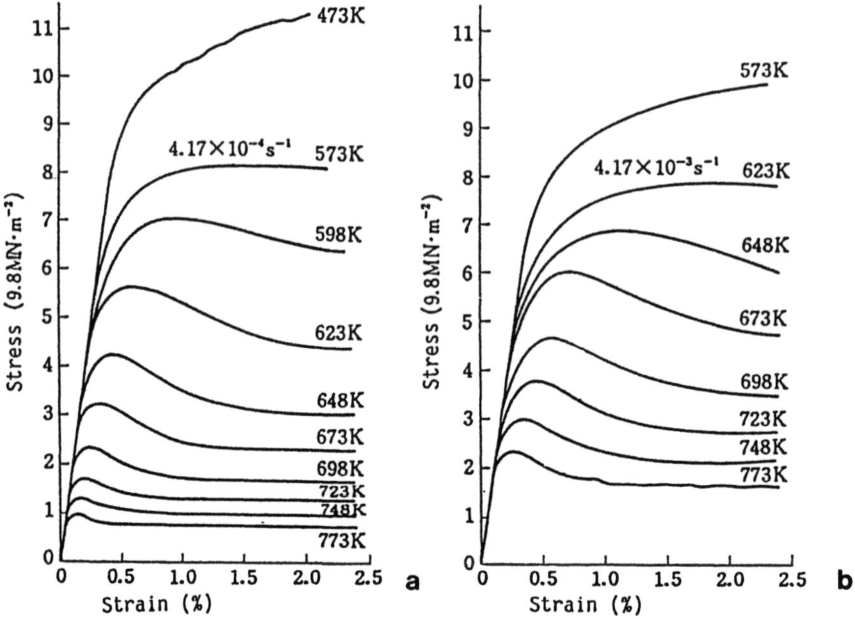

Fig. 9.32a,b. Temperature dependence of yielding behavior of an Al-5.5 at.% Mg alloy [9.63]. Tensile strain rate of (a) $4.17 \times 10^{-4}\,\mathrm{s}^{-1}$, (b) $4.17 \times 10^{-3}\,\mathrm{s}^{-1}$

be connected with the fact that ϱ_u also increases with τ and the deformation mechanism changes to the intermediate temperature type when τ is too high [9.79].

Figure 9.32 shows the early stage of stress-strain curves. The work softening does not start at the onset of plastic deformation, but a very high work-hardening is observed prior to reaching the peak stress. However, this is not a true work hardening, but an apparent one arising from the too low plastic strain rate compared with the imposed apparent strain rate. In a tensile test, the apparent strain rate $\dot{\varepsilon}_a$ containing elastic deformation, is kept constant. Equation (8.57) shows that the apparent work-hardening rate, $d\sigma/d\varepsilon_a$, becomes high when the plastic strain rate is very low in the initial stage of deformation; apparent work hardening occurs [9.68]. As the plastic strain rate gradually increases with the increase in stress from 0, the apparent work hardening rate decreases from the very high value at the beginning.

c) Stress Exponent of Alloys and the Origin of Internal Stress. Assuming that the flow stress τ is equal to the drag stress τ_d, and the relation of (8.14) holds between τ and dislocation density, *J. Weertman* and *J.R. Weertman* [9.80] derived the relation that $\dot{\gamma}$ is proportional to τ^3. Although this is the simplest theory for deriving the 3rd-power law, neither of the two assumptions made by them seems appropriate.

194

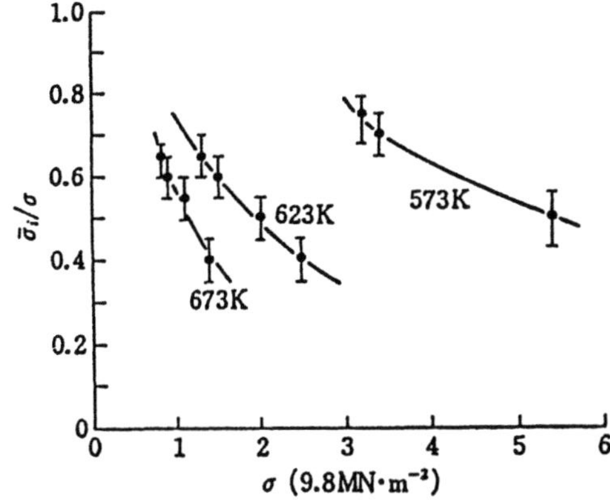

Fig. 9.33. Dependence of internal stress contribution $\bar{\sigma}_i/\sigma$ on flow stress σ in steady state deformation of an Al-5.7at.%Mg alloy [9.16]

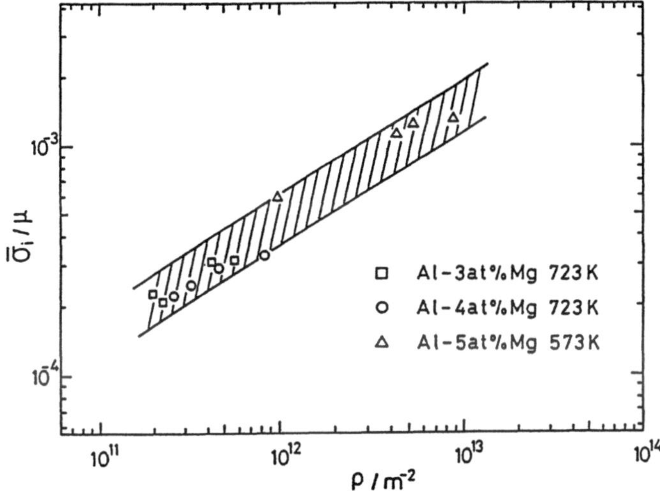

Fig. 9.34. Relationship between dislocation density ϱ and modulus-compensated internal stress $\bar{\sigma}_i/\mu$ determined in steady-state deformation of Al-Mg alloys [9.82]

The assumption of $\tau = \tau_d$ means that the contribution of internal stress to the flow stress is negligible compared to that of the drag stress τ_d. According to the experimental results (Fig. 9.33) obtained by *Toma* et al. [9.16], however, the contribution of internal stress cannot be neglected, even in the alloys where the 3rd-power law holds. The internal stresses shown in the figure were measured by using the stress dip and extrapolation technique. *Yoshinaga* et al. [9.81] have obtained almost the same results by the stress change technique. Figure 9.34 [9.82] shows the dislocation density dependence of internal stress, where the density was measured by TEM observation after cooling under load. This dependence agrees well with the attractive junction theory [9.83].

In solution-hardened alloys, the dislocations belonging to various slip systems are considered to move simultaneously at velocities proportional to their effective

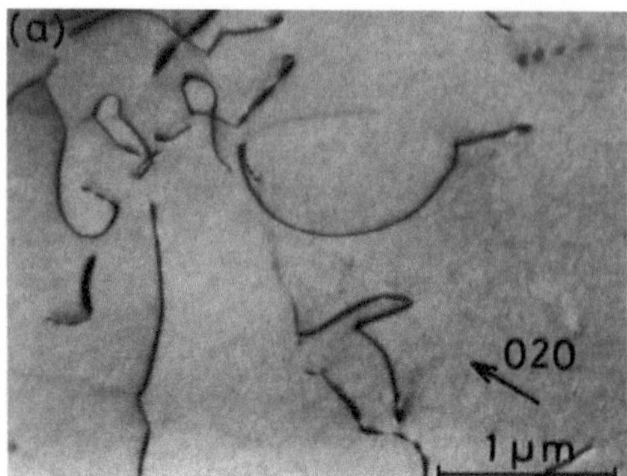

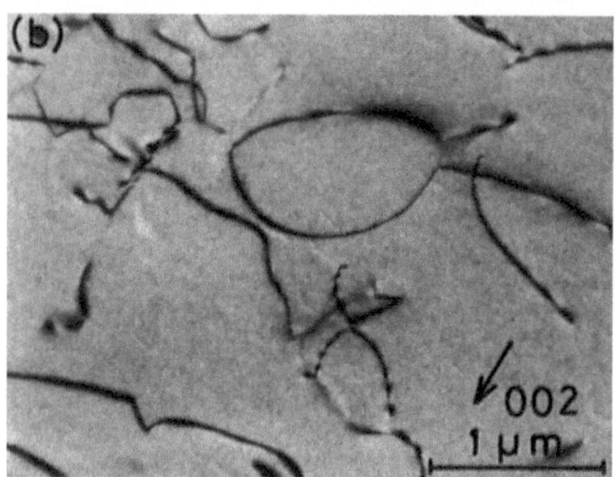

Fig. 9.35a,b. Dislocation structure observed in an Al-4.56 at.% Mg alloy crept well into the steady state at 623 K under 30.5 MN m^{-2} [9.82] (a) 020 reflection. (b) 002 reflection

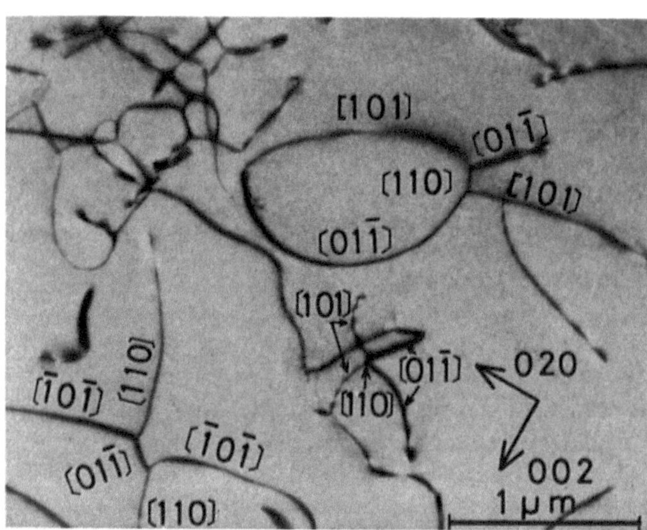

Fig. 9.36. Dislocation junctions visualized by superposing Fig. 35a and b [9.82]

resolved shear stresses as predicted by (9.53). Thus many slip systems should be activated simultaneously even in single crystals. *Otsuka* and *Horiuchi* [9.84] verified this simultaneous activation by observing the lattice reorientation of Al-3 at.% Mg alloy single crystals deformed under tension at 632 K. The moving dislocation belonging to different slip systems act as forest dislocations to one another, and thus there are many chances to form junctions. Figure 9.35 shows the attractive junctions thus formed. The TEM images were observed by using (a) 020 reflection, and (b) 002 reflection. By superposing (a) and (b), we can see all kinds of $\langle 110 \rangle$-type dislocations. Figure 9.36 is the superposed image, where many junctions are observed. The indices in the figure show the direction of the Burgers vector of each dislocation segment. Using this superposing technique, *Hayakawa* et al. [9.82] recently identified that most dislocation arcs such as shown in Fig. 9.18 have junctions at the ends. The observations show that the internal stress arises from attractive junctions.

Mills et al. [9.85] reported different experimental results and proposed a dislocation loop model. According to them, a transient deformation acoompanying a stress reduction is mainly due to the rapid motion of screw parts of dislocation loops and true internal stress is much lower than the drag stress for edge dislocations. However, the TEM observations described above show that perfect loops rarely exist and most of the loop-like dislocations are actually arcs having both ends at junctions.

The second assumption made by *J. Weertman* and *J.R. Weertman* is also hardly acceptable, because τ given by $\alpha\mu b\sqrt{\varrho}$ is not the drag stress τ_d but the internal stress τ_i. It seems that in their assumption there is some confusion of τ_d and τ_i.

Takeuchi and *Argon* [9.86] derived the 3rd-power law more reasonably from the multiplication and annihilation rates of dislocations. However, there is a problem in that they also assumed $\tau = \tau_d$. Although they theoretically showed that when only a single slip system operates for deformation, τ is approximately equal to τ_d, multiple slip systems are inevitably activated in solution-hardened alloys as described above.

Oikawa et al. [9.87] showed experimentally that in Orowan's equation, where $\dot{\gamma} = \varrho b\bar{v} = 2\varrho b\bar{\tau}_e/B$, $\bar{\tau}_e \propto \tau^{1.39}$ in Al-Mg alloys. Combining this with the experimental relation obtained by *Horiuchi* and *Otsuka* [9.64], $\varrho \propto \tau^{1.67}$, they showed that the stress exponent of $\dot{\gamma}$ becomes $1.39 + 1.67 = 3.06 \cong 3$. According to these experimental results, the 3rd power law is no more than an approximate relation arising from the stress dependence of ϱ and $\bar{\tau}_e$.

Once the relation between ϱ and τ in the steady state has been established, the relation between τ and $\dot{\gamma}$ is known from $\dot{\gamma} = 2\varrho bB(\tau - \alpha\mu b\sqrt{\varrho})$. *Yoshinaga* et. al. [9.88] have shown that if $d\varrho/d\gamma$ is proportional to τ [9.89], the relation between ϱ and τ obtained experimentally by *Horiuchi* and *Otsuka* can be derived.

10. High-Temperature Deformation Mechanism in Composite Materials

Intermetallic compounds and ceramics such as oxides, carbides, nitrides, etc., are usually high in strength but brittle, whereas metals are in general ductile but low in strength. Many efforts are being made to produce composite materials having both high strength and ductility by combining them. Even in composite materials the strength decreases at high temperatures in many cases. The decrease is also caused by the increase in diffusion velocity. In this chapter will be described the kinds of materials which exist as useful composites and how their deformation is affected by diffusion.

10.1 Types of Composite Materials

Figure 10.1 illustrates three types of composite materials: (a) dispersion-strengthened materials, where hard particles (rarely soft particles) are distributed, (b) fiber-reinforced materials, where strong fibers are included, and (c) Lamella-reinforced materials, where hard lamellas are included layer by layer. The hard phase is ideally zero-dimensional in (a), one-dimensional in (b) and two-dimensional in (c). As the number of dimensions increases, the constraint on the dislocation motion increases in the soft phase. At high temperatures, cross slip and climbing of dislocations easily occur, and particles in (a) are easily surmounted by dislocations if the interaction of particles with dislocations is repulsive. Dislocations cannot pass through the fibers in (b) without cutting them or leaving loops behind around the fibers. In the case of (c), the movement of dislocations is necessarily hindered by the phase boundaries, except for dislocations moving parallel to the lamellas. Thus, it is considerd that as the dimension of the hard phase increases, the strength increases at the expense of the ductility.

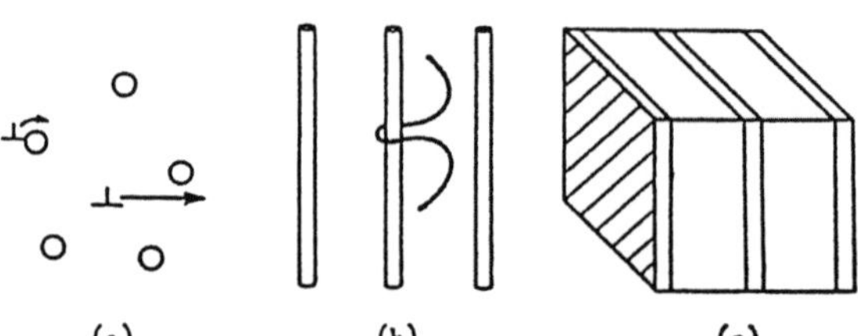

Fig. 10.1a–c. Various types of composite structure

If the soft phase is surrounded three-dimensionally by the hard phase, the constraint on the soft phase is perfect and the ductility may decrease greatly. This type of composite is difficult to produce, and zero-, one- or two-dimensional composite materials are mostly being investigated as the method of strengthening metals. Especially dispersion strengthened materials are being widely studied, because such materials are easily produced by powder metallurgy, for example.

10.2 High-Temperature Deformation Mechanism in Dispersion-Strengthened Materials

As the mechanism for the passage of dislocations through the dispersed particles, the climb model has long been widely accepted. This model can well explain the decrease in yield stress at high temperatures. Recently, however, a model of attractive interaction between dislocations and hard particles was proposed. The model predicts a high threshold stress for creep deformation which does not decrease at high temperature, if the stress is normalized by the elastic modulus. In the following, the models and the related experimental observations will be described in historical order.

10.2.1 Climb Model

If a glide dislocation cannot change its slip plane by cross slip or climb, it must bulge out between the dispersed particles and then pass through them, leaving dislocation loops behind around the particles, as schematically shown in Fig. 10.2. This passing mechanism is called the *Orowan mechanism* [10.1], and the loops left behind are called *Orowan loops*. In this case, the resistance to dislocation motion is given by

$$\tau_{Or} = \frac{2\alpha T_d}{\lambda b} \tag{10.1}$$

and is called the *Orowan stress*. Here, λ is the spacing between particles, T_d is the line tension of a dislocation, and α a constant depending on the distribution state of the particles (0.82 for a random distribution [10.2, 3]). The Orowan stress is a athermal stress of the kind called internal stress in this text.

Even when cross slip occurs locally at particles, the resistance will not decrease, because superjogs having a height of about the particle diameter should be produced [10.4]. However, at high temperatures where dislocations are able to climb, the resistance should decrease.

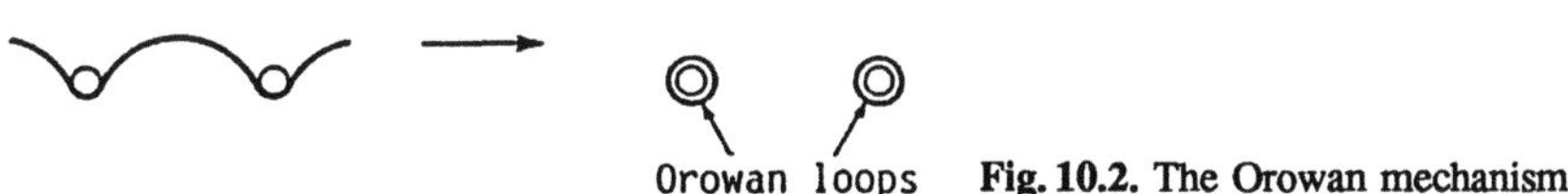

Fig. 10.2. The Orowan mechanism

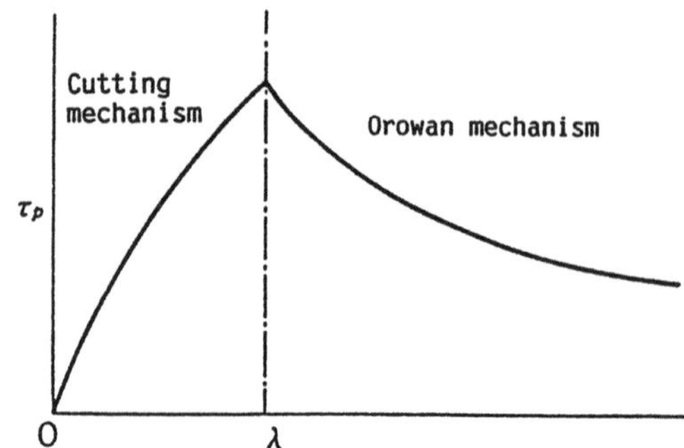

Fig. 10.3. Schematic representation of the relationship between inter-particle spacing λ and stress τ_p required for a dislocation to pass through the dispersed particles without cross slip or climb

A particle of strength F, which is lower than $2T_d$, will be cut by the dislocation when the angle φ made by two branches of the dislocation pinned at the particle reaches $2\cos^{-1}(F/2T_d)$. This passing mechanism is called the cutting mechanism. Since the particle strength increases as the particle size increases, there is a critical size above which the Orowan mechanism operates and below which the cutting mechanism operates. For a constant volume fraction of dispersed particles, the Orowan stress increases as the particle size decreases, because the interparticle spacing, λ, decreases. However, when the size decreases further and enters the cutting region, the resistance decreases as λ decreases, because F decreases along with the particle size. As a result, it is expected that the resistance depends on λ as shown in Fig. 10.3 [10.5]. At high temperatures, on the other hand, the dislocation can pass the particles even under a stress lower than the Orowan value not by cutting but by climbing. For simplification, let us assume here that the particles are too hard to be cut, and consider how the deformation occurs at high temperatures below and above the Orowan stress.

Ansell and *Weertman* [10.6] derived the strain rate $\dot{\gamma}$ as follows. When the applied stress is lower than the Orowan stress and deformation is controlled by the rate of emission of dislocation loops,

$$\dot{\gamma} = M\pi L^2 b \frac{dn_d}{dt} . \tag{10.2}$$

Here, M is the dislocation-source density, πL^2 the area swept by a dislocation loop moved from the source to the place of pair annihilation, and dn_d/dt the rate of emission per source.

When H is the distance between the slip planes of the two oppositely signed dislocations that are to be pair-annihilated, there is a source in every volume $\pi L^2 H$. However, *Ansell* and *Weertman* assumed $M/3 \cong 1/\pi L^2 H$, by considering that the screw dislocation should be easily pair annihilated by cross slip and the edge dislocation may control the deformation. Then,

$$ML^2 \cong \frac{1}{H} . \tag{10.3}$$

Here, H was assumed to be two times as large as the particle diameter d_v.

In a steady state, the multiplication rate of dislocations is equal to the annihilation rate. When the climb velocity of dislocations is v_{cl}, the time required for pair annihilation is d_v/v_{cl}. As the force acting on a dislocation of length b is τb^2, the climbing rate is given by

$$v_{cl} = \frac{D}{k_B T}\tau b^2 .$$ (10.4)

Then,

$$\frac{dn_d}{dt} = \frac{1}{d_v/v_{cl}} = \frac{\tau b^2 D}{d_v k_B T} .$$ (10.5)

Putting (10.3) and (10.5) in (10.2), we obtain

$$\dot{\gamma} = \frac{\pi \tau b^3 D}{2 k_B T d_v^2} .$$ (10.6)

Above the Orowan stress, dislocations move leaving loops behind. When the back stress from the Orowan loops thus formed balances the applied stress, the deformation stops at low temperature. At high temperatures, however, the loops can disappear by climbing and the deformation proceeds. Thus, the process of disappearence of loops controls the deformation. In a steady state, the production of loops balances their disappearence.

The number n of dislocations that can pile up in a spacing λ between particles, is

$$n \cong \frac{2\tau \lambda}{\mu b} ,$$ (10.7)

and the stress on the leading dislocation of the pile-up is $n\tau b$ per unit length. When the back stress from the existing loops becomes lower than $n\tau b$, a new dislocation loop is formed. Since the back stress from an existing loop which has climbed by h is $\mu b^2/h$, by equating this with $n\tau b$ we obtain the climb distance required for a new loop formation as

$$h = \frac{\mu b}{n\tau} = \frac{\mu^2 b^2}{2\tau^2 \lambda} .$$ (10.8)

A force of $n\tau b^2$ acts on the loop per atomic length. Then the climbing rate is given by

$$v_{cl} = \frac{D}{k_B T} n\tau b^2 = \frac{2\tau^2 b\lambda D}{\mu k_B T} ,$$ (10.9)

and

$$\frac{dn_d}{dt} = \frac{1}{h/v_{cl}} = \frac{4\tau^4 \lambda^2 D}{b\mu^3 k_B T} .$$ (10.10)

From (10.2, 3, 10) and $H = 2d_v$, we obtain

$$\dot{\gamma} = \frac{2\pi \tau^4 \lambda^2 D}{d_v \mu^2 k_B T} .$$ (10.11)

As seen in (10.2), the above theory neglects the effect of the time required for dislocations to move to their disappearing places. Under a low stress below the Orowan stress, after the leading dislocation is annihilated, the next dislocation should sweep an area of nearly πL^2 before it is blocked, because the dislocation pile-up is, if any, very small, under this low-stress condition. Therefore, the dislocation must climb over the particles existing in this area, and the time required to sweep the area may be much longer than that for the leading dislocation to climb by d_v. As a result, this theory probably overestimates the creep rate below the Orowan stress. In fact, this theory gives a creep rate 10^3–10^4 times higher than the experimental rate [10.7]. Further, the theory predicts that the stress exponent of the steady-state creep rate should be 1. However, the experimentally obtained exponents are much larger than 1, nearly 4, except in the report by *Blickensderfer* [10.7].

On the other hand, *Brown* and *Ham* [10.4] considered that the process of local climbing of dislocations at the particles is rate controlling below the Orowan stress. This is called the local climb mechanism.

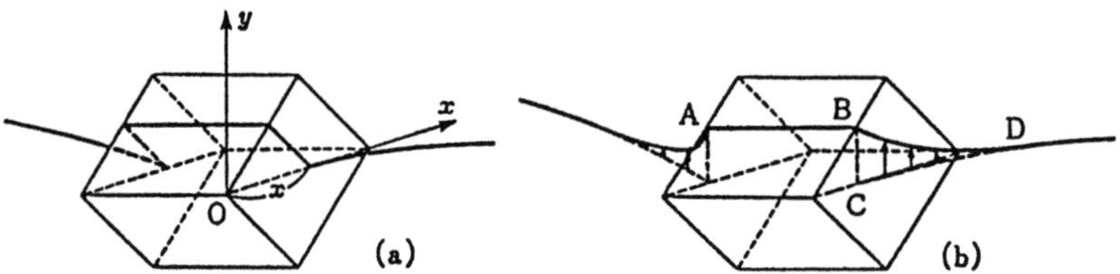

Fig. 10.4a,b. Climb mechanisms. (a) Local climb. (b) Unravelling, which leads to general climb in this extreme case

For convenience of analysis, it is assumed that the particles are a cube with a side of length d and the slip plane is parallel to the cube's diagonal plane. If the dislocation climbs sticking to the particle surface as illustrated in Fig. 10.4a, the minimum length of the dislocation line moved from O by x is $d+2(x/\sqrt{2})$. When the dislocation moves further by δx keeping the minimum length condition, the length increases by $\sqrt{2}\delta x$ and the energy of the dislocation line increases by $\sqrt{2}T_d\delta x$. On the other hand, the applied stress does work $\tau b\lambda\delta x$, where λ is the spacing between the particles. Then, the change in total energy is given by

$$\delta E = (\sqrt{2}T_d - \tau b\lambda)\delta x \ . \tag{10.12}$$

The number of vacancies required for the dislocation to climb by δy is

$$\delta n_v = \frac{bd\delta y}{\Omega} \ , \tag{10.13}$$

where Ω is the volume of a vacancy and $\delta y = \delta x/2$ in the model of Fig. 10.4a. The energy per vacancy is thus given by

$$\varepsilon = \frac{\delta E}{\delta n_{\mathrm{v}}} = \frac{2\Omega}{bd} \left(\sqrt{2}T_{\mathrm{d}} - \tau b\lambda \right) , \tag{10.14}$$

and the vacancy concentration at the climbing dislocation is given by

$$c_{\mathrm{v}} = c_{\mathrm{ve}} \exp \left(-\frac{\varepsilon}{k_{\mathrm{B}}T} \right) .$$

Shewfelt and *Brown* [10.8] considered that diffusion on the particle surface is much faster than lattice diffusion and assumed that the vacancy concentration is thus uniform along the surface and it is the equilibrium concentration, c_{ve}, sufficiently far away from the particle. On these assumptions, they solved the diffusion equation and showed that the vacancy flow at the surface is given by

$$J = \frac{2\pi d D_{\mathrm{l}}}{\Omega} \left[\exp \left(-\frac{\varepsilon}{k_{\mathrm{B}}T} \right) - 1 \right] \cong \frac{4\pi D_{\mathrm{l}}}{b k_{\mathrm{B}}T} \left(\tau b\lambda - \sqrt{2}T_{\mathrm{d}} \right) , \tag{10.15}$$

Where D_{l} is the coefficient of lattice self-diffusion. Then the dislocation velocity is given by

$$\frac{dx}{dt} = \frac{2dy}{dt} = 2\frac{Jb}{d/b} = \frac{8\pi b D_{\mathrm{l}}}{d k_{\mathrm{B}}T} \left(\tau b\lambda - \sqrt{2}T_{\mathrm{d}} \right) . \tag{10.16}$$

Based on the local climb model, *Shewfelt* and *Brown* [10.8] calculated the flow stress for random particle distribution by computer simulation and showed that the flow stress can decrease to a level of about 0.4 times the Orowan stress, and the flow stress for a given ϱ_{m} is proportional to the logarithm of the strain rate. Their theoretical result agrees well with their experimental results of temperature and strain-rate dependences of yield stress.

The above two theories consider in common that the rate-determining process is the climbing of dislocations over the particle, though one considers that the process is dislocation annihilation, while the other considers that the process is dislocation glide. According to the theory of *Ansell* and *Weertman*, the source density M affects both the multiplication rate of dislocations and the area swept by a dislocation until annihilation, but the two effects cancel each other to give no effect on the strain rate. This means that the deformation of a dispersion-strengthened material should proceed in its steady state from the very beginning. This prediction, however, goes against the experimental fact that the material shows a normal transient creep. According to the theory of *Shewfelt* and *Brown*, on the other hand, the observed normal transient indicates that the mobile dislocation density ϱ_{m} decreases in the transient stage. If ϱ_{m} really decreases, the immobile dislocation density should be increased, because the total dislocation density is increased by deformation. This leads to an increase in internal stress, which reduces the effective stress assisting the climbing of dislocations over particles. Therefore, the effect of an increase in internal stress should be involved in the theory for steady-state deformation.

Since work hardening occurs in the transient stage, *Lagneborg* [10.10] considered that the rate-determining step is not the multiplication process of dislocations, namely the work-hardening process, but the recovery process. First, he [10.11] modified the model of *Brown* and *Ham* for the deformation below the Orowan stress, as is shown in Fig. 10.4b. Namely, he pointed out that unravelling of the dislocation from the particle should occur by climbing of the part of dislocation near the particle, because the dislocation must bend sharply to stick to the particle and the dislocation line tension serves as the driving force of the unravelling. He studied how the resistance to the dislocation motion is reduced by this unravelling effect. As the length of the part climbing away from the slip plane (CD in the figure) increases, the increment of dislocation length required for surmoumting the particle decreases. As a result, the resistance decreases. When the applied stress τ is reduced, the force $\tau b\lambda$ required for the dislocation segment on the particle (AB in the figure) to climb decreases, thus reducing the climb velocity. Then the climb velocity due to the line-tension increases relatively, giving rise to an increase in the length of the part CD. In this way, the resistance by particles τ_p decreases with a decrease of τ, and the stress dependence of τ_p is given by

$$\tau_\mathrm{p} \cong 0.7\tau \ . \tag{10.17}$$

Lagneborg and *Bergman* [10.5] assumed that the flow stress is given by the sum of τ_p and the resistance arising from the interaction of dislocations,

$$\bullet\tau = \tau_\mathrm{p} + \alpha\mu b\sqrt{\varrho} \ , \tag{10.18}$$

and derived a relation between the flow stress and strain rate in the steady state.

The method of estimation of ϱ is the same as described in Sect. 9.1.2. The rate of increase of dislocation density is given from (9.14) by

$$\dot{\varrho}^{+} = \dot{\gamma}\frac{\partial\varrho}{\partial\gamma} = \frac{\dot{\gamma}}{sb} \tag{10.19}$$

and the rate of decrease is given from (9.7 and 10a) by

$$\dot{\varrho}^{-} = 2M_\mathrm{d}T_\mathrm{d}\varrho^{2} \ . \tag{10.20}$$

Here, s is the mean free path of dislocations, and M_d the dislocation mobility. According to *Friedel*, $M_\mathrm{d} = 2bc_jD/k_\mathrm{B}T$, see (9.10a). Then, in the steady state of $\dot{\varrho}^{+} = \dot{\varrho}^{-}$,

$$\dot{\gamma} = 2bsM_\mathrm{d}T_\mathrm{d}\varrho^{2} \ . \tag{10.21}$$

A problem that is different from pure metals is the effect of dispersed particles on the network growth. *Lagneborg* and *Bergman* evaluated the effect by considering that τ_p also acts as resistance to the growth. In the growth, the effective stress for dislocation links to surmuount the particles is $(\tau - \tau_\mathrm{p})\lambda b$. Assuming here that τ_p is given by (10.17) and τ is given by the stress $T_d/l_1 b$ arising from the line tension of the dislocation, the driving force for a dislocation to climb over a particle becomes $0.3(\lambda/d)T_d/l_1$, where d is the particle diameter. When the volume fraction of the dispersed particles is f, then $\lambda/d \cong 1/\sqrt{f}$. Thus the climb velocity becomes

$$v_{cl} = 0.3 f^{-1/2} M_d \frac{T_d}{l_1} , \qquad (10.22)$$

which is $0.3 f^{-1/2}$ times as large as the climb velocity in pure metals. Therefore, when $0.3 f^{-1/2} \gg 1$, or $f \ll 0.09$, the climb velocity at the particle is very high compared with the velocity in the absence of particles, so that the presence of particles has a negligible effect on the network growth. Since $f = 0.01-0.02$ in many dispersion-strengthened materials, the effect of the particles is negligible.

From (10.18 and 21) we obtain

$$\dot{\gamma} = 2bs M_d T_d \left(\frac{\tau - \tau_p}{\alpha \mu b} \right)^4 . \qquad (10.23)$$

By showing that many experimental results are as shown in Fig. 10.5, *Lagneborg* and *Bergman* emphasized the validity of (10.23). Below the Orowan stress (more exactly below $1/0.7$ of the Orowan stress), τ_p is proportional to τ as shown in (10.17) and therfore $\dot{\gamma}$ is proportional to τ^4, which agrees well with the stress exponent in the lower stress range of the figure. When 0.7τ is increased over τ_{Or}, $\tau_p = \tau_{Or}$, for any value of τ. Then the apparent stress exponent n' increases suddenly to $\tau/(\tau - \tau_{Or})$ times the true stress exponent n, as shown in (9.18). This prediction also agrees well with the experimental results shown in the figure. If this is the case, the factor n'/n should decrease toward 1 as τ is increased further. However, the figure does not show this decrease in n'. This is probably because the experimental stress range is too small for the decrease to be observed. On the other hand, as seen in Fig. 10.6, the experimental results obtained by *Lund* and *Nix* [10.12] using Ni-20% Cr single crystals containing about 2 vol.% of ThO_2 show that n' really decreases toward 4 as the creep stress is increased above the Orowan stress. The observed stress exponent ranges from 4 to 40. This remarkable change in stress exponent depending on the value of τ/τ_{Or} is well explained by the theory.

Either of the theories described above predicts the equality of the activation energies for deformation and self-diffusion. However, the experimental value is frequently larger than the activation energy for lattice self-diffusion; sometimes several times as large as that energy. To account for this disagreement various theories such as grain boundary dislocation-source theory [10.6] and interstitial-type jog-drag theory [10.13] have been proposed so far. Among them, the explanation proposed by *Lund* and *Nix* [10.14] seems most reasonable as shown below.

Whenever the high activation energy was obtained, the apparent stress exponent n' was large and the energy was obtained from the temperature dependence of creep rate under the constant stress condition. In other words, the activation

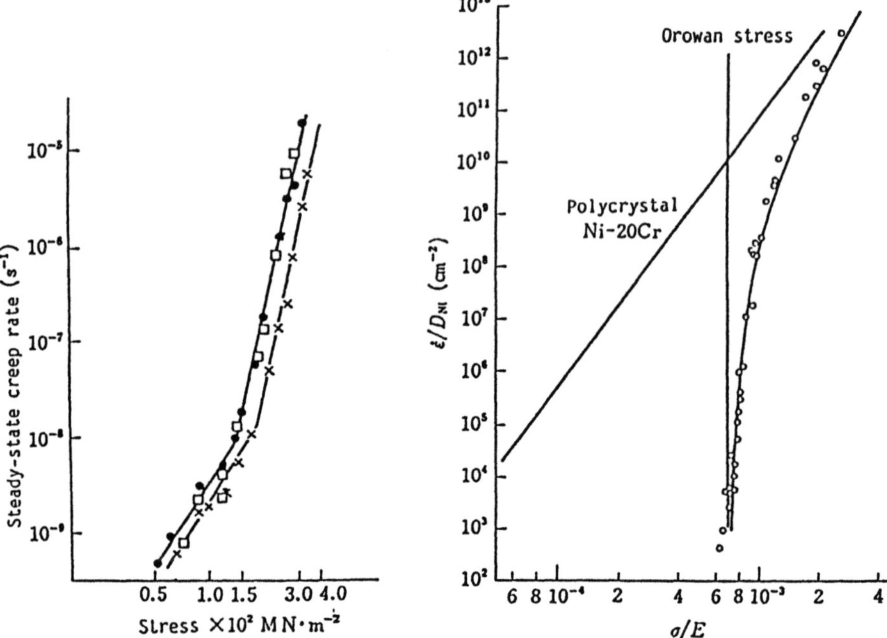

Fig. 10.5. Stress dependence of creep rate in nickel-based dispersion-hardened alloys [10.5]. □ Ni-19.3 Cr-0.87 Al-2.06 Ti ● Ni-18.4 Cr-2.6 Al × Ni-19.3 Cr-0.73 Al-2.66 Ti

Fig. 10.6. Modulus-compensated stress dependence of creep rate compensated by the diffusion coefficient of Ni in Ni-20 Cr-2 ThO₂ single crystals [10.12]

energy Q_c was estimated by assuming that the steady-state creep rate is given by $\dot{\varepsilon} = A_1\sigma^{n'} \exp(-Q_c/RT)$. More correctly, however, the creep rate should be expressed as

$$\dot{\varepsilon} = A \left(\frac{\sigma}{E(T)}\right)^{n'} \exp\left(-\frac{Q_c}{RT}\right) , \tag{10.24}$$

where $E(T)$ is the elastic modulus depending on temperature. The apparent activation energy obtained by neglecting the temperature dependence of $E(T)$ is given by

$$Q_{app} = -R \left(\frac{\partial \ln \dot{\varepsilon}}{\partial(1/T)}\right)_\sigma .$$

Thus the true activation energy is related to Q_{app} by

$$Q_c = -R \left(\frac{\partial \ln \dot{\varepsilon}}{\partial(1/T)}\right)_{\sigma/E} = Q_{app} + n'R\frac{T^2}{E}\frac{dE}{dT} . \tag{10.25}$$

Since $dE/dT < 0$, Q_{app} can be much larger than Q_c for a large n'. In fact, *Lund* and *Nix* [10.14] showed that once this correction is made, Q_c becomes approximately equal to the activation energy for self-diffusion.

206

10.2.2 Attractive Interaction

In the preceding section, the unravelling of a dislocation from a particle was shown to decrease the strengthening effect of dispersed particles. In contrast to the local climb model (Fig. 10.4a), this by-pass mechanism is called the general climb [10.15], because all parts of a dislocation climb away from its original slip plane in the extreme case. According to *Lagneborg* [10.10], no threshold stress for creep deformation exists, as seen from (10.17). Against his theory, however, a threshold stress should exist, because some parts of a dislocation climb up and other parts climb down from its original slip plane, with the consequence that some extension of dislocation line length is required for the dislocation to by-pass the particles. However, the threshold stress for general climb is very low compared with the Orowan stress [10.15] as long as the volume fraction of hard particles is not very large [10.16].

On the other hand, a threshold stress as high as the Orowan stress is sometimes observed. Figure 10.7 [10.17] is an example. To explain the high threshold stress, *Srolovitz* et al. [10.18] proposed a model of diffusionally modified elastic interaction. When the dispersed particles are harder than the matrix, the image force repels the dislocation from the particle. This is the case at low tempera-

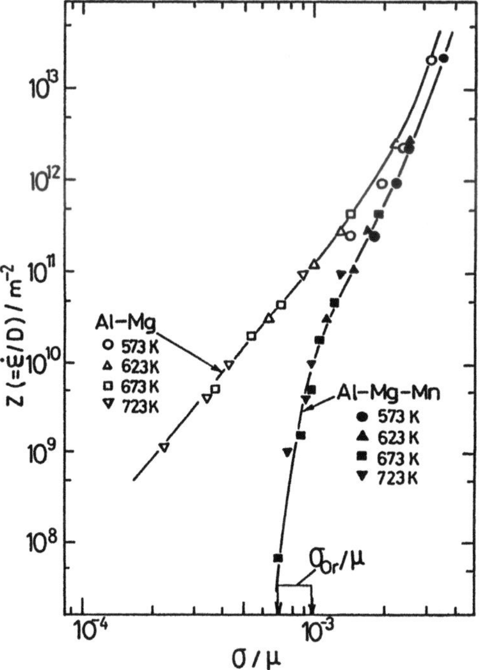

Fig. 10.7. Relationship between modulus-compensated stress σ/μ and Zener-Holomon parameter Z (strain rate compensated by diffusion coefficient) for steady-state deformation of Al-3 at.% Mg (solid solution) and Al-3 at.% Mg-1 at.% Mn (Al-3 at.% Mg solid solution hardened by Al_6Mn particle dispersion [10.17])

tures. At high temperatures, however, the shear stress imposed by the dislocation on the interface between the particle and the matrix can be relaxed by the interface sliding, and the normal stress can also be relaxed by lattice diffusion. Then the elastic interaction between the dislocation and the particle becomes attractive. The relaxation at the interface may easily occur for incoherent particles, while for coherent particles it is thought to occur less easily and then the climb mechanism should operate. Using this model, *Srolovitz* et al. showed that the threshold stress is expected to be as high as the Orowan stress. If the relaxation is complete, the particle is the same as a void with respect to the interaction with the dislocation, and the dislocation core spreads when the dislocation comes into contact with the particle. In this case, the void-strengthening theory proposed by *Weeks* et al. [10.19] may be applied to the dispersion hardening. *Horiuchi* [10.20] showed that the void-strengthening theory can well be applied to alloys hardened by liquid particles. Such an attractive interaction was first observed by *Schröder* and *Arzt* [10.21] for hard particles. However, a clear dislocation contrast was observed along the dislocation segment stuck to a particle. This means that the relaxation was not complete. Then *Arzt* and *Wilkinson* [10.22] calculated the effect of the degree of relaxation on the threshold stress and showed that a rather weak attractive interaction gives a threshold stress higher than that for the climb mechanism.

Figure 10.8 shows an almost complete relaxation recently observed by *Nakashima* et al. [10.17], where the dislocation contrast is hardly observed around particles. This interaction really gives a threshold stress as high as the Orowan stress, as shown in Fig. 10.7.

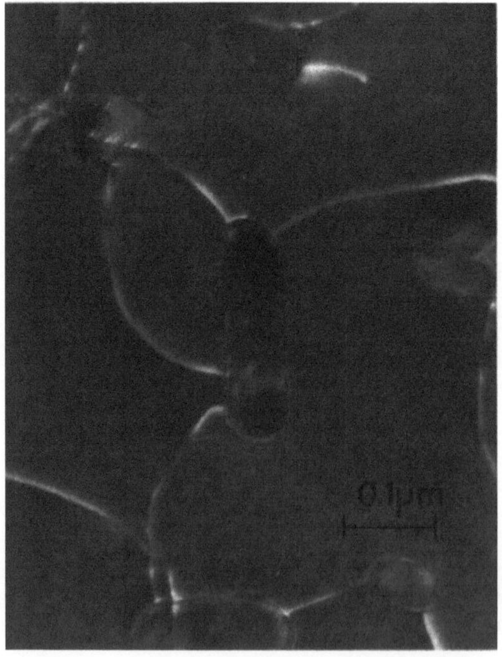

Fig. 10.8. Weak-beam TEM image in Al-3 at.% Mg-1 at.%Mn with Al$_6$Mn particles that has crept well into the steady state at 573 K under 57.5 MN m^{-2} and cooled rapidly under load [10.17]

10.3 High-Temperature Deformation Mechanism in Fiber- and Lamella-Reinforced Materials

The deformation of the soft phase is necessarily obstructed by the hard phase when $\varphi = 0$ (parallel) or $\pi/2$ (perpendicular), because the slip plane intersects the hard phase, where φ is the angle between the stress axis and the direction of the hard phase. When φ is intermediate, the slip deformation parallel to the hard phase is not hindered. However, as other slip deformations inclined to the hard phase are obstructed, a strengthening effect is still expected.

When the shear deformation of the soft phase is completely hindered, the hard phase will be deformed by nearly the same amount as the soft phase as long as the dislocations stopped by the hard phase are not absorbed at the interface. As the applied stress is gradually increased, at first the soft phase begins to deform plastically, while the hard phase continues to deform elastically. Only after the hard phase yields or fractures does a large plastic deformation occur in the composite material. Even when the hard phase is brittle, a noticeable fracture of the hard phase does not occur in many cases until the deformation exceeds about 0.5%. Therefore, the strength of composite materials is generally much higher than that of matrix metals.

In the case of Al-CuAl$_2$ (θ phase) eutectic lamellar composites, where the strength of the hard phase is much higher than that of the soft matrix and the volume fraction f is as high as 50%, the yield strength of the composite is almost completely determined by the strength of the hard phase, and the soft phase offers almost no contribution to the strength of the composite; the soft phase simply plays the role of a stress-transmitting medium [10.23]. In such a case, the strength of the composite is almost independent of lamella spacing λ [10.23]. On the other hand, as is the case of Mo-TiC eutectic lamellar composites, where f is about 25% and φ is random, the contribution to the strength of the metal matrix sometimes exceeds 50% of the 0.2% proof stress of the composite, because the work hardening in the soft phase is increased by the hard phase obstructing the deformation [10.24]. In the former case, the deformation behavior of the composite is not much different from that of the single phase material of the hard phase [10.25]. On the other hand, in the latter case, the strength of the composite is strongly affected by the recovery, because a significant part of the strength is carried by the work hardening of the soft phase.

Denoting the lamella spacing by λ, the mean free path of dislocations is approximately $\sqrt{2}\lambda/2$ and the increase in dislocation density due to the tensile or compressive plastic deformation ε_p may be given by

$$\Delta\varrho = 2^{3/2}\frac{\varepsilon_p}{b\lambda} , \tag{10.26}$$

because $\varepsilon_p \cong \gamma_p/2 \cong (\Delta\varrho b\lambda/\sqrt{2})/2$. Then the increase in internal stress due to dislocation interaction is given by

$$\sigma_i \cong 2\tau_i = 2\alpha\mu b\sqrt{\Delta\varrho} \, .$$

Here, the effect of initial dislocation density is neglected, assuming that the density is small compared with $\Delta\varrho$. If the effective stress is σ_e, the flow stress of the soft phase is given by

$$\sigma_M = \sigma_e + 2^{7/4}\alpha\mu b \left(\frac{b\varepsilon_p}{\lambda}\right)^{1/2} \tag{10.27}$$

in the case of no recovery effect.

On the other hand, the elastic strain of the soft phase is $\varepsilon_{eM} = \sigma_M/E_M$, where E_M is the Young's modulus of that phase. Assuming that the hard phase of Young's modulus E_H is deformed elastically by the same amount as the soft phase ($\varepsilon_{eM} + \varepsilon_p$), the stress acting on the hard phase should be

$$\sigma_H = E_H(\varepsilon_{eM} + \varepsilon_p) = E_H\left(\frac{\sigma_M}{E_M} + \varepsilon_p\right) \, . \tag{10.28}$$

Therefore, the flow stress of the composite with a volume fraction f of hard phase is given by

$$\sigma_c = \sigma_M(1 - f) + \sigma_H f = \sigma_M \left[1 + \left(\frac{E_H}{E_M} - 1\right)f\right] + E_H f \varepsilon_p \, . \tag{10.29}$$

Conversely, by measuring σ_c the flow stress of the soft phase can be estimated from

$$\sigma_M = \frac{\sigma_c - E_H f \varepsilon_p}{1 + (E_H/E_M - 1)f} \, . \tag{10.30}$$

When the elastic strain of the hard phase ε_{eH} is smaller than $\varepsilon_{eM} + \varepsilon_p$, (10.28) leads to an overestimation of σ_H and (10.30) leads to an underestimation of σ_M.

Kurishita et al. [10.24] showed that for a Mo-TiC lamellar composite with $\lambda = 0.25\,\mu$m and $f = 0.26$, the σ_M estimated by (10.30) reaches no less than 30 times the 0.2% proof stress of Mo single-phase polycrystals at $\varepsilon_p = 0.002$. Since this can be an under- but not an overestimation as mentioned above, it is known that the effect of work hardening is very large.

On the other hand, the theoretical σ_M obtained from (10.27) was lower by about 36% than that obtained by putting the experimental σ_c at room temperature into (10.30). Because no recovery is considered to occur at room temperature, the difference indicates that (10.27) tends to underestimate σ_M. The reason for this may be that the theory neglects the effects of (1) dislocation pile-up, (2) the image force from the hard phase and (3) the decrease in mean free path due to the increase in dislocation density.

As the temperature rises, the strength of Mo-TiC composites decreases rapidly as shown in Fig. 10.9. Although TiC can be deformed plastically above 1100 K [10.26, 27], the plastic deformation up to 0.2% proof stress of the composite occurs mainly in the Mo phase [10.24]. Accordingly, the temperature dependence

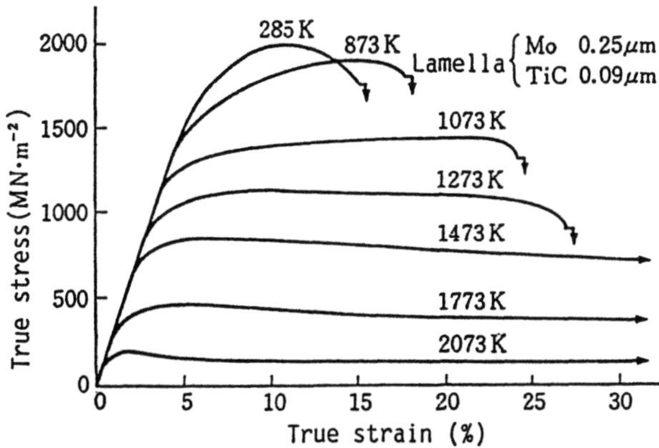

Fig. 10.9. Stress-strain curves of a Mo-TiC eutectic lamellar composite [10.24]

of 0.2% proof stress of the composite is considered to be a reflection of the recovery effect on the work hardening in the soft phase. In fact, the ratio of σ_M in the composite at $\varepsilon_p = 0.002$ to the 0.2% proof stress of the Mo single-phase material decreases rapidly as the temperature rises, from 30 at room temperature to 6 at 1773 K. The ratio becomes less than 1 at 2073 K. This incomprehensible result may arise from the fact that the TiC phase also yields or that dislocations enter the interface and disappear there at this high temperature, because the σ_M is calculated on the assumption that the TiC phase is deformed only elastically and the strain matching is complete between the two phases. If dislocations in the Mo phase disappear at the interface, the total strength (elastic + plastic) in the Mo phase can become more than that in the TiC phase.

The dislocation density in the soft phase can be decreased by the network growth or pair annihilation, whereas the dislocations stopped by the interface cannot disappear unless they are absorbed by the interface or they cut the hard phase. The ease of absorption may depend not only on temperature (through interface diffusion) but also on the image force, caused by the difference in elastic modulus, and the coherency of the interface (through the stress relaxation at the interface). Unfortunately study of this problem is not advanced. However, the observation described in Sect. 10.2.2 shows that an incoherent interface easily absorbs dislocations at high temperatures.

If dislocations are annihilated at the interface, the strain in the soft phase can be much larger than that in the hard phase. In fact, it is known that when the Mo-TiC composite is compressed at high temperatures, the soft phase is squeezed out to cover the surface of the specimen [10.24].

Another type of composite material is that where the component phases are both ductile. Though the Mo-TiC described above also belongs to this type at high temperature and high strain, typical examples are metal-metal composites. It is especially interesting to know whether or not some composite effect really exists even when the strengths of both phases are not very different. When the orientation or the crystal structure is different between the component phases, it is known without doubt that the interface obstructs dislocation motion in the same manner as the grain boundary in a single-phase material. However, the Ag-Cu eutectic lamellar composite made by *Cline* and *Stein* [10.28] using the unidirectional solidification technique has the same orientation as well as the same fcc crystal structure in both phases. In this case, therefore, the obstructing effect of the interface on dislocation motion might be weak, though the difference in lattice constant may offer some effect. Nevertheless, it was reported that the strength of the composite is higher than those of the single phase component crystals and that the 0.5% proof stress increases inversely proportional to $\sqrt{\lambda}$; for $\lambda = 0.1\,\mu$m the proof stress is no less than 4 times as high as the strength calculated from the strengths of single-phase crystals by using the rule of mixtures, see (10.29). In addition, the strength decreases little at high temperatures, maintaining about one half of the strength at room temperature even at high temperatures (100 K below the melting point). According to the analysis of *Cline* and *Stein* the high strength can well be explained by the image force acting to resist dislocation motion. The weak temperature dependence of the proof stress suggests that the coherency of the interface is so high that the effect of dislocation absorption at the interface is negligible. Since the image force is proportional to the difference in elastic modulus between the two phases, this kind of composite effect is considered to increase as the difference increases as long as the absorption effect is negligible, and since the temperature dependence of the elastic modulus is weak, the composite effect is temperature insensitive.

References

Chapter 1

1.1 W.T. Read, Jr.: *Dislocations in Crystals* (McGraw-Hill, New York 1953)
1.2 A.H. Cottrell: *Dislocations and Plastic Flow in Crystals* (Oxford University Press, London 1953)
1.3 H. Suzuki: *Introduction to Dislocation Theory* (Agne, Tokyo 1967), in Japanese
1.4 J. Friedel: *Dislocations* (Pergamon, Oxford 1964)
1.5 F.R.N. Nabarro: *Theory of Crystal Dislocations* (Oxford University Press, London 1967)
1.6 J.P. Hirth, J. Lothe: *Theory of Dislocations*, 2nd ed. (Wiley, New York 1982)
1.7 F.R.N. Nabarro (ed.): *Dislocation in Solids*, Vols. 1–7 (North-Holland, Amsterdam 1979–1987)

Chapter 2

2.1 P. Hirsch: Can. J. Phys. **45**, 663 (1967)
2.2 S.J. Basinski, Z.S. Basinski: In *Dislocations in Solids*, Vol. 4, ed. by F.R.N. Nabarro (North-Holland, Amsterdam, 1979) p.261
2.3 R. Peierls: Proc. Phys. Soc. **52**, 34 (1940)
2.4 F.R.N. Nabarro: Proc. Phys. Soc. **59**, 256 (1947)
2.5 W.T. Brydges: Philos. Mag. **15**, 1079 (1967)
2.6 G. Saada: Acta Metall. **8**, 84 (1960)
2.7 T. Suzuki: In *Dislocation Dynamics*, ed. by A.R. Rosenfield, G.T. Hahn, A.L. Bement, Jr., R.I. Jaffee (McGraw-Hill, New York 1956/57) p.551
2.8 T. Suzuki, T. Ishii: In Proc. Int. Conf. on Strength of Metals and Alloys, Trans. Jpn. Inst. Met. **9**, 687 (1967/68)
2.9 T. Suzuki, T. Ishii: In *Physics of Strength and Plasticity*, ed. by A.S. Argon (MIT Press, Cambridge, MA 1969) p.159
2.10 W.G. Johnston, J.J. Gilman: J. Appl. Phys. **30**, 129 (1959)
2.11 W.G. Johnston, J.J. Gilman: Solid State Phys. **13**, 147 (1962)
2.12 C.S. Pang, J.M. Galligan: Phys. Rev. Lett. **43**, 1595 (1979)
2.13 K. Marukawa: J. Phys. Soc. Jpn. **22**, 499 (1967)
2.14 H. Suga, T. Imura: Jpn. J. Appl. Phys. **12**, 751 (1973)
2.15 M. Kleintges, R. Labush, H. Brion, P. Haasen: Acta Metall. **25**, 1247 (1977)
2.16 H. Ney, R. Labush, P. Haasen: Acta Metall. **25**, 1257 (1977)
2.17 O.P. Pope, T. Vreeland, Jr., D.S. Wood: J. Appl. Phys. **38**, 3595 (1967)
2.18 H. Saka, K. Noda, T. Imura: Cryst. Lattice Defects **4**, 45 (1973)
2.19 T. Suzuki, H. Kojima: Acta Metall. **14**, 913 (1966)
2.20 H.W. Shadler: Acta Metall. **12**, 861 (1964)
2.21 D.F. Stein, J.R. Low: J. Appl. Phys. **31**, 362 (1960)
2.22 A.J.E. Foreman, M.J. Makin: Philos. Mag. **13**, 911 (1966)
2.23 P. Wynblatt: In *Rate Processes in Plastic Deformation of Materials*, ed. by J.C.M. Li, A.K. Mukherjee (A.S.M., Metals Park, OH 1975) p.156

2.24   J.D. Eshelby, F.C. Frank, F.R.N. Nabarro: Philos. Mag. **42**, 351 (1951)
2.25   J.M. Galligan: Phys. Acoust. **16**, 173 (1982)
2.26   A.V. Granato, K. Lücke: J. Appl. Phys. **27**, 583 (1956)
2.27   R.W. Balluffi, A.V. Granato: In *Dislocations in Solids*, Vol. 4, ed. by F.R.N. Nabarro (North-Holland, Amsterdam 1979) Chap. 1
2.28   A. Hikata, R.A. Johnston, C. Elbaum: Phys. Rev. B **2**, 4856 (1970)
2.29   A. Hikata, C. Elbaum: Trans. Jpn. Inst. Met. Suppl. **9**, 46 (1968)
2.30   O.M. Mitchell: J. Appl. Phys. **36**, 2083 (1965)
2.31   R.A. Moog: Ph.D. Thesis, Cornell University (1965)
2.32   F. Fanti, J. Holder, A.V. Granato: J. Acoust. Soc. **45**, 279 (1969)
2.33   T. Suzuki, A. Ikushima, M. Aoki: Acta Metall. **12**, 1231 (1964)
2.34   W.P. Mason, A. Rosenberg: Phys. Rev. **151**, 434 (1966)
2.35   G.A. Alers, D.O. Thompson: J. Appl. Phys. **32**, 283 (1961)
2.36   R.H. Stein, A.V. Granato: Acta Metall. **10**, 358 (1962)
2.37   T. Holstein: see the paper by Tittman and Bömmel: Phys. Rev. **151**, 740 (1966)
2.38   V. Ya. Kravchenko: Sov. Phys.–Solid State **8**, 740 (1966)
2.39   A.D. Brailsford: Phys. Rev. **186**, 959 (1969)
2.40   C. Kittel: *Quantum Theory of Solids*, (Wiley, New York 1963) Chap. 17
2.41   A. Seeger, H. Engelke: In *Dislocation Dynamics*, ed. by A.R. Rosenfield, G.T. Hahn, A.L. Bement, Jr., R.I. Jaffee (McGraw-Hill, New York 1968) p. 623
2.42   P.P. Gruner: In *Fundamental Aspects of Dislocation Theory*, Vol. 1, ed. by J.A. Simmons, R. de Wit, R. Bullough (National Bureau of Standards, Washington, DC 1970) p. 363
2.43   A.D. Brailsford: J. Appl. Phys. **43**, 1380 (1972)
2.44   Y. Kogure, Y. Hiki: J. Phys. Soc. Jpn. **38**, 471 (1975)
2.45   T. Ninomiya: J. Phys. Soc. Jpn. **25**, 830 (1968)
2.46   T. Ninomiya: In *Fundamental Aspects of Dislocation Theory*, Vol. 1, ed. by J.A. Simmons, R. de Wit, R. Bullough (National Bureau of Standards, Washington, DC 1970) p. 315
2.47   T. Ninomiya: J. Phys. Soc. Jpn. **36**, 399 (1974)
2.48   A.V. Granato: Phys. Rev. **111**, 740 (1958)
2.49   J.A. Garber, A.V. Granato: J. Phys. Chem. Solids **31**, 1863 (1970)
2.50   K. Ohashi, Y.H. Ohashi: Philos. Mag. A **38**, 187 (1978)

Chapter 3

3.1   J. Weertman: *Physics and Chemistry of Ice*, ed. by E. Whalley, S.J. Jones, L.W. Gold (Royal Society of Canada, Ottawa 1973)
3.2   J.D. Eshelby: Philos. Mag. **6**, 953 (1961)
3.3   P. Schiller: Z. Phys. **153**, 1 (1958)
3.4   T. Suzuki: In *Dislocation Dynamics*, ed. by A.R. Rosenfield, G.T. Hahn, A.L. Bement, Jr., R.I. Jaffee (McGraw-Hill, New York 1968) p. 551
3.5   T. Suzuki, T. Ishii: Proc. Int. Conf. on Strength of Metals and Alloys, Trans. Jpn. Inst. Met. **9**, Suppl., 687 (1967/68)
3.6   A.A. Hendrickson, M.E. Fine: Trans. AIME **221**, 103 (1961)
3.7   Z.S. Basinski, D. Dove: See S.J. Basinski, Z.S. Basinski: In *Dislocations in Solids*, Vol. 4, ed. by F.R.N. Nabarro (North-Holland, Amsterdam 1979) p. 261
3.8   T. Suzuki, Jpn. J. Appl. Phys. **20**, 449 (1981)
3.9   K. Kamada, I. Yoshizawa: J. Phys. Soc. Jpn. **31**, 1056 (1971)
3.10   N. Büttner, E. Nembach: Z. Metallkd. **76**, 82 (1984)
3.11   H. Suga, Y. Sueki, M. Higuchi, T. Imura: Jpn. J. Appl. Phys. **14**, 379 (1976)
3.12   F. Iida, T. Suzuki, E. Kuramoto, S. Takeuchi: Acta Metall. **27**, 637 (1978)
3.13   T. Kan, P. Haasen: Z. Metallkd. **61**, 355 (1970)
3.14   P. Jax, P. Kratochvil, P. Haasen: Acta Metall. **18**, 237 (1970)
3.15   G. Kostorz, P. Haasen: Z. Metallkd. **60**, 26 (1969)

3.16 R.A. Kloske, M.E. Fine: Trans. AIME **245**, 217 (1969)
3.17 P. Haasen: In *Dislocations in Solids*, Vol. 4, ed. by F.R.N. Nabarro (Noth-Holland, Amsterdam 1979), p.155
3.18 H. Suzuki: In *Dislocations and Mechanical Properties of Crystals*, ed. by J.C. Fisher, W.G. Johnston, R. Thomson, T. Vreeland, Jr. (Wiley, New York 1956/57) p.361
3.19 R.L. Fleischer: Acta Metall. **11**, 203 (1963)
3.20 J. Friedel: *Dislocations* (Pergamon, New York 1964)
3.21 T. Suzuki, T. Ishii: In *Physics of Strength and Plasticity*, ed. by A.S. Argon (MIT Press, Cambridge, MA 1969) p.687
3.22 N.F. Mott, F.R.N. Nabarro: *Strength of Solids* (Phys. Soc., London 1947) p.1
3.23 R. Labusch: Phys. Status Solidi **41**, 659 (1970)
3.24 B.R. Riddhagni, R.M. Asimov: J. Appl. Phys. **39**, 4144 (1968)
3.25 R. Labusch, J.S. Ahearn, G. Grange, P. Haasen: *Rate Processes in Plastic Deformation*, AIME Symp. Series (Plenum, New York 1975)
3.26 R.W. Balluffi, A.V. Granato: In *Dislocations in Solids*, Vol.4, ed by F.R.N. Nabarro (North-Holland, Amsterdam 1979) p.1
3.27 R.B. Schwarz, R. Labusch: J. Appl. Phys. **49**, 5174 (1978)
3.28 U.F. Kocks, A.S. Argon, M.F. Ashby: *Thermodynamics and Kinetics of Slip* (Pergamon, New York 1975)
3.29 A.J.E. Foreman, M.J. Makin: Philos. Mag. **14**, 911 (1966)
3.30 T. Suzuki: Jpn. J. Appl. Phys. **20**, 449 (1981)
3.31 E. Orowan: *Symposium on Internal Stresses* (Inst. Metals, London 1948) p.451
3.32 R.L. Fleischer: Acta Metall. **9**, 996 (1961)
3.33 F.R.N. Nabarro:*Theory of Crystal Dislocation* (Clarendon, Oxford 1967)
3.34 A. Seeger, P. Haasen: Philos. Mag. **3**, 470 (1958)
3.35 J. Saxl: Cz. J. Phys. B **14**, 381 (1964)
3.36 F.R.N. Nabarro: In *The Physics of Metals*, Vol.2, ed. by P.B. Hirsch (Cambridge University Press, Cambridge 1975) Chap.4
3.37 S. Takeuchi: Scr. Metall. **2**, 481 (1968)

Chapter 4

4.1 H. Kojima, T. Suzuki: Phys. Rev. Lett. **21**, 896 (1968)
4.2 V.V. Pustovalov, V.I. Startsev, V.S. Fomenko: Sov.Phys.–Solid State **11**, 1119 (1969)
4.3 M. Suenaga, J.M. Galligan: Phys. Acoust. **9**, 1 (1972)
4.4 J.M. Galligan: Phys. Acoust. **16**, 173 (1982)
4.5 G. Kostorz: Phys. Status Solidi B **58**, 9 (1973)
4.6 F.R.N. Nabarro, A.T. Quintanilha: In *Dislocations in Solids*, Vol.5, ed. by F.R.N. Nabarro (North-Holland, Amsterdam 1980) p.193
4.7 V.I. Startsev: In *Dislocations in Solids*, Vol.6, ed. by F.R.N. Nabarro (North-Holland, Amsterdam 1983) p.143
4.8 T. Suzuki: Jpn. J. Appl. Phys. **20**, 449 (1981)
4.9 A.V. Granato: Phys. Rev. B **4**, 2196 (1971)
4.10 M. Suenaga, J.M. Galligan: Scr. Metall. **5**, 829 (1971)
4.11 F. Iida, T. Suzuki, E. Kuramoto, S. Takeuchi: Acta Metall. **27**, 637 (1978)
4.12 E. Kuramoto, F. Iida, S. Takeuchi, T. Suzuki: J. Phys. Soc. Jpn. **38**, 431 (1975)
4.13 J. Bardeen, L. Cooper, J. Schrieffer: Phys. Rev. **108**, 1175 (1957)
4.14 A.D. Brailsford: Phys. Rev. **186**, 959 (1969)
4.15 G. Kostorz, Acta Metall. **21**, 167 (1973)
4.16 G.A. Alers, O. Buck, B.R. Tittman: Phys. Rev. Lett. **23**, 290 (1969)
4.17 J.M. Galligan, J.H. Tregilgas: Scr. Metall. **9**, 1321 (1975)
4.18 H. Kojima, T. Moriya, T. Suzuki: J. Phys. Soc. Jpn. **38**, 1032 (1975)
4.19 T. Suzuki: *Rate Processes in Plastic Deformation* (American Society for Metals, Metals Park, OH 1976) p.249

4.20 G. Kostorz: J. Low. Temp. Phys. **10**, 167 (1973)
4.21 G. Kostorz: Philos. Mag. **27**, 634 (1973)
4.22 T. Suzuki, T. Ishii: In *Physics of Strength and Plasticity*, ed. by A.S. Argon (MIT Press Cambridge, MA 1969) p.687
4.23 Z.S. Basinski, R.A. Foxall, R. Pascual: Scr. Metall. **6**, 807 (1972)
4.24 K. Kamada, I. Yoshizawa: J. Phys. Soc. Jpn. **31**, 1056 (1971)
4.25 N. Büttner, E. Nembach: Z. Metallkd. **76**, 82 (1984)
4.26 H. Suga, Y. Sueki, M. Higuchi, T. Imura: Jpn. J. Appl. Phys. **14**, 379 (1976)
4.27 R.B. Schwarz, R.D. Isaac, A.V. Granato: Phys. Rev. Lett. **38**, 554 (1977)
4.28 V.S. Natsik: Sov. Phys.–JETP **34**, 1359 (1972)
4.29 A.V. Granato: In *Internal Friction and Ultrasonic Attenuation in Solids*, ed. by R.R. Hashiguchi (Tokyo University Press, Tokyo 1977) p.81
4.30 G.P. Huffman, N.P. Louat: Phys. Rev. Lett. **24**, 1055 (1970)
4.31 M.I. Kaganov, V.D. Natsik: Sov. Phys.–JETP Lett. **11**, 379 (1970)
4.32 T. Moriya, T. Suzuki: J. Phys. Soc. Jpn. **56**, 3941 (1987)

Chapter 5

5.1   R.E. Peierls: Proc. Phys. Soc. London **52**, 34 (1940)
5.2   F.R.N. Nabarro: *Theory of Crystal Dislocations* (Clarendon Press, Oxford 1967) p.175
5.3   A. Seeger: Philos. Mag. **1**, 651 (1956)
5.4   A. Seeger, P. Schiller: Acta Metall. **10**, 348 (1962)
5.5   J.P. Hirth, J. Lothe: *Theory of Dislocations* (McGraw-Hill, New York 1968)
5.6   J.E. Dorn, S. Rajnak: Trans. Met. Soc. AIME **230**, 1052 (1964)
5.7   V. Celli, M. Kabler, T. Ninomiya, R. Thomson: Phys. Rev. **131**, 58 (1963)
5.8   P. Guyot, J.E. Dorn: Can. J. Phys. **45**, 983 (1967)
5.9   T. Mori, M. Kato: Philos. Mag. A **43**, 1315 (1981)
5.10 R.I. Arsenault: Acta Metall. **15**, 501 (1967)
5.11 K. Ono, A.W. Sommer: Metall. Trans. **1**, 877 (1970)
5.12 A. Sato, M. Meshii: Acta Metall. **21**, 753 (1973)
5.13 Y. Kawata, S. Ishioka: Philos. Mag. A **48**, 921 (1983)
5.14 J. Lothe, J.P. Hirth: Phys. Rev. **115**, 543 (1959)
5.15 A. Seeger, P. Schiller: In *Physical Acoustics*, Vol. III A, ed. by W.P. Mason (Academic, New York 1966) p.361
5.16 J. Castaing, P. Veyssiere, L.P. Kubin, J. Rabier: Philos. Mag. A **44**, 1407 (1981)
5.17 F. Louchet: Philos. Mag. A **43**, 1289 (1981)

Chapter 6

6.1   H. Suzuki: *Dislocation Dynamics*, ed. by A.R. Rosenfield, G.T. Hahn, A.L. Bement, Jr., R.I. Jaffee (McGraw-Hill, New York 1968) P. 679
6.2   For research up to 1973, see V. Vitek: Cryst. Lattice Defects **5**, 1 (1974)
6.3   Z.S. Basinski, M.S. Duesbery, R. Taylor: Can. J. Phys. **49**, 2160 (1971)
6.4   V. Vitek: Proc. R. Soc. London A **352**, 109 (1976)
6.5   V. Vitek, M. Yamaguchi: J. Phys. F **3**, 537 (1973)
6.6   R.A. Johnson: Phys. Rev. A **134**, 1329 (1964)
6.7   H. Suzuki: In *Fundamental Aspects of Dislocation Theory*, Vol. I, ed by J.A. Simons, R. de Wit, R. Bullough (National Bureau of Standards, Spec. Publ. 317, 1970) p.253
6.8   F. Minami, E. Kuramoto, S. Takeuchi: Phys. Status Solidi A **22**, 81 (1974)
6.9   E. Kuramoto, F. Minami, S. Takeuchi: Phys. Status Solidi A **22**, 411 (1974)
6.10 S. Takeuchi: *Interatomic Potentials and Crystalline Defects*, ed. by J.K. Lee (The Metals Society AIME, Warrendale, PA 1981) p.201

6.11 V. Vitek, R.C. Perrin, D.K. Bowen: Philos. Mag. **21**, 1049 (1970)
6.12 V. Vitek: *Strength of Metals and Alloys* (American Society of Metals, Metals Park, OH 1970) p.389
6.13 A. Seeger, C. Wüthrich: Nuovo Cimento B **33**, 38 (1976)
6.14 S. Takeuchi: Philos. Mag. A **39**, 661 (1979)
6.15 A. Sato, K. Masuda: Philos. Mag. B **43**, 1 (1981)
6.16 E. Kuramoto, Y. Aono, T. Tsutsumi: In *Strength of Metals and Alloys*, Vol.1, ed. by R.C. Gifkins (Pergamon, Oxford 1982) p.69
6.17 F. Minami, E. Kuramoto, S. Takeuchi: Phys. Status Solidi A **12**, 581 (1972)
6.18 M.S. Duesbery, V. Vitek, D.K. Bowen: Proc. R. Soc. London A **332**, 85 (1973)
6.19 M.S. Duesbery: Proc. R. Soc. London A **392**, 145, 175 (1984)
6.20 T. Taoka, S. Takeuchi, E. Furubayashi: J. Phys. Soc. Jpn. **19**, 701 (1964)
6.21 S. Takeuchi, E. Kuramoto: J. Phys. Soc. Jpn. **38**, 480 (1975)
6.22 A.J.E. Foreman, M.A. Jaswon, J.K. Wood: Proc. Phys. Soc. London A **64**, 156 (1951)
6.23 R. Conte, P. Groh, B. Escaig: Phys. Status Solidi **28**, 475 (1968)
6.24 K. Kitajima, Y. Aono, H. Abe, E. Kuramoto: Scr. Metall. **13**, 1033 (1979)
6.25 Z.S. Basinski:Proc. R. Soc. London A **240**, 229 (1957)
6.26 K. Kitajima, Y. Aono, H. Abe, E. Kuramoto: In *Strength of Metals and Alloys*, ed. by P. Hassen, V. Gerold, G. Kostorz (Pergamon, Oxford 1979) p.965
6.27 K. Kitajima, Y. Aono, E. Kuramoto: Scr. Metall. **15**, 919 (1981)
6.28 S. Takeuchi, E. Kuramoto, T. Suzuki: Acta Metall. **20**, 909 (1972)
6.29 K. Murakami, Y. Umakoshi, M. Yamaguchi: Philos. Mag. **37**, 719 (1978)
6.30 S. Takeuchi, T. Hashimoto, K. Suzuki, M. Ichihara: Acta Metall. **30**, 513 (1982)
6.31 H. Saka, G. Taylor, Philos. Mag. A **43**, 1377 (1981)
6.32 S. Takeuchi: Philos. Mag. A **41**, 541 (1980)
6.33 H. Conrad: *The Relation Between the Structure and Mechanical Properties of Metals* (H.M.S.O., London 1963) p.476
6.34 Y. Aono, E. Kuramoto, K. Kitajima: Rep.Res. Inst. Appl. Mechanics, Kyushu Univ. **29** (92), 127 (1981)
6.35 S. Takeuchi, K. Maeda: Acta Metall. **25**, 1485 (1977)
6.36 S. Takeuchi, T. Hashimoto, K. Maeda: Trans. Jpn. Inst. Metals **23**, 60 (1982)
6.37 Y. Aono, K. Kitajima, E. Kuramoto: Scr. Metall. **15**, 275 (1981)
6.38 A. Seeger: Z. Metallkd. **72**, 369 (1981)
6.39 Z.S. Basinski, M.S. Duesbery, G.S. Murty: Acta Metall. **29**, 801 (1981)
6.40 A.J. Garrat-Read, G. Taylor: Philos. Mag. **33**, 577 (1976)
6.41 R. Creten, J. Bressers, P. De Messter: Mater. Sci. Eng. **19**, 51 (1977)
6.42 A.J. Garrat-Read, G. Taylor: Philos. Mag. **39**, 597 (1979)
6.43 F. Louchet, L.P. Kubin: Acta Metall. **23**, 19 (1975)
6.44 H. Matsui, H. Kimura: Mater. Sci. Engl. **24**, 247 (1976)
6.45 V.K. Sethi, R. Gibara: Scr. Metall. **9**, 527 (1975)
6.46 K. Kojima, M. Meshii: Phys. Status Solidi A **39**, 491 (1977)
6.47 R.I. Arsenault: Acta Metall. **15**, 501 (1967)
6.48 K. Ono, A.W. Sommer: Metall. Trans. **1**, 877 (1970)
6.49 A. Sato, M. Meshii: Acta Metall. **21**, 753 (1973)
6.50 J. Weertman: J. Appl. Phys. **12**, 1685 (1958)
6.51 H. Suzuki: In *Dislocations in Solids*, Vol. 4, ed. by F.R.N. Nabarro (North-Holland, Amsterdam 1979), p.191
6.52 S. Takeuchi, H. Yoshida, T. Taoka: Trans. Jpn. Inst. Met. **9**, Suppl., 715 (1968)
      S. Takeuchi: J. Phys. Soc. Jpn. **27**, 929 (1969)
6.53 T.E. Mitchell, R.L. Raffo: Can. J. Phys. **45**, 1047 (1967)
6.54 H. Suzuki: In *The Structure and Properties of Crystal Defects*, ed. by V. Paidar, L.Lejcek (Elsevier, Amsterdam 1984), p.205
6.55 Y. Aono, K. Kitajima, E. Kuramoto: Scr. Metall. **14**, 321 (1980)

Chapter 7

7.1   D.J.H. Cockayne, A. Hons: J. de Phys. **40**, C6–11 (1979)

7.2   A. Olsen, J.C.H. Spence: Philos. Mag. A **43**, 945 (1981)

7.3   W.T. Read, Jr.: Philos. Mag. **45**, 775 (1954)

7.4   R. Labusch, W. Schröter: In *Dislocations in Solids*, Vol. 5, ed. by F.R.N. Nabarro (North-Holland, Amsterdam 1980) p. 127

7.5   V.V. Kveder, Yu.A. Osip'yan, W. Schröter, G. Zoth: Phys. Status Solidi A **72**, 701 (1982)

7.6   H. Ono, K. Sumino: J. Appl. Phys. **54**, 4426 (1983)

7.7   E. Weber, H. Alexander: J. de Phys. **40**, C6–101 (1979)

7.8   V.A. Grazhulis, V.V. Kveder, Yu. A. Osip'yan, Y.H. Lee, R.L. Kleinhenz, H. van Camp, C.P. Sholes, J.W. Corbett: Phys. Lett. A **66**, 398 (1978)

7.9   M. Suezawa, K. Sumino, M. Iwaizumi: *Defects and Radiation Effects in Semiconductors 1980*, Inst. Phys. Conf. Ser. No. 59 (Institute of Physics, Bristol 1981) p. 407

7.10 H. Alexander, C. Kisielowski-Kemmerich, E.R. Weber: Physica B **116**, 583 (1983)

7.11 P.B. Hirsch: J. de Phys. **40**, C6–27 (1979)

7.12 R. Jones: J. de Phys. **40**, C6–33 (1979)

7.13 S. Marklund: Phys. Status Solidi B **92**, 83 (1979); ibid. **100**, 77 (1980)

7.14 J.R. Patel, A.R. Chaudhuri: Phys. Rev. **143**, 601 (1966)

7.15 M. Heggie, R. Jones: Philos. Mag. B **48**, 365 (1983)

7.16 S. Schäfer: Phys. Status Solidi **19**, 297 (1967)

7.17 J.R. Patel, P.E. Freeland: J. Appl. Phys. **42**, 3298 (1971)

7.18 For a review on the dislocation mobility in semiconductors see F. Louchet, A. George: J. de Phys. **44**, C4–51 (1983)

7.19 A.R. Chaudhuri, J.R. Patel, L.G. Rubin: J. Appl. Phys. **33**, 2736 (1962)

7.20 T. Suzuki, H. Kojima: Acta Metall. **14**, 913 (1966)

7.21 H.L. Frisch, J.R. Patel: Phys. Rev. Lett. **18**, 784 (1967)

7.22 V.N. Erofeev, V.I. Nikitenko: Sov. Phys. – JETP **60**, 1780 (1971)

7.23 A. George, C. Escaravage, G. Champier, W. Schröter: Phys. Status Solidi B **53**, 483 (1972)

7.24 J.R. Patel, L.R. Testardi, P.E. Freeland: Phys. Rev. B **13**, 3548 (1976)

7.25 J.R. Patel, L.R. Testardi: Appl. Phys. Lett. **30**, 3 (1977)

7.26 M. Imai, K. Sumino: Philos. Mag. **47**, 599 (1983)

7.27 K. Wessel, H. Alexander: Philos. Mag. **35**, 1523 (1977)

7.28 H. Alexander, C. Kisielowski-Kemmerich, E.R. Weber: Physica B **116**, 583 (1983)

7.29 M. Mihara, T. Ninomiya: J. Phys. Soc. Jpn. **25**, 19 (1968)

7.30 H. Steinhardt, S. Schäfer: Acta Metall. **19**, 65 (1971)

7.31 S.K. Choi, M. Mihara: J. Phys. Soc. Jpn. **32**, 1154 (1972)

7.32 S.A. Erofeeva, Yu. A. Osip'yan: Sov. Phys. – Solid State **15**, 583 (1973); ibid. **16**, 2076 (1975)

7.33 V.B. Osvenskii, L.P. Kholodnyi: Sov. Phys. – Solid State **14**, 2822 (1973)

7.34 S.K. Choi, M. Mihara, T. Ninomiya: Jpn. J. Appl. Phys. **16**, 737 (1977)

7.35 S.K. Choi, M. Mihara, T. Ninomiya: Jpn. J. Appl. Phys. **17**, 329 (1978)

7.36 T. Ninomiya: J. de Phys. **40**, C6–143 (1979)

7.37 K. Suzuki, M. Ichihara, S. Takeuchi, K. Nakagawa, M. Maeda, H. Iwanaga: Philos. Mag. A **49**, 451 (1984)

7.38 H.-J. Möller: Acta Metall. **26**, 963 (1978)

7.39 F. Louchet: Philos. Mag. A **43**, 1289 (1981)

7.40 F. Louchet: *Microscopy of Semiconducting Materials 1981*, Inst. Phys. Conf. Ser., No. 60 (Institute of Physics, Bristol 1981) p. 35

7.41 P.B. Hirsch, A. Ourmazd, P. Pirouz: *Microscopy of Semiconducting Materials, 1981*, Inst. Phys. Conf. Ser., No. 60 (Institute of Physics, Bristol 1981) p. 29

7.42 P. Haasen: J. de Phys. **40**, C6–111 (1979)

7.43 P.B. Hirsch: J. de Phys. **40**, C6–117 (1979)

7.44 R. Jones: Philos. Mag. B **42**, 213 (1980)

7.45 K. Maeda: Proc. Yamada Conf. on Dislocations in Solids (University of Tokyo Press, Tokyo 1985) p. 425
7.46 E.g.: G.C. Kuczynki, K.R. Iyer, C.W. Allen: J. Appl. Phys. **43**, 1337 (1972)
7.47 P.N. Petroff: J. de Phys. **40**, C6–201 (1979)
7.48 M. Iwamoto, A. Kasami: Appl. Phys. Lett. **28**, 591 (1976)
7.49 B. Monemar, G.R. Woolhouse: Appl. Phys. Lett. **29**, 605 (1976)
7.50 K. Ishida, T. Kamejima, J. Matsui: Appl. Phys. Lett. **31**, 397 (1977)
7.51 H. Nakanishi, S. Kishino, N. Chinone, R. Ito: J. Appl. Phys. **48**, 2771 (1979)
7.52 K. Maeda, S. Takeuchi: Jpn. J. Appl. Phys. **20**, L165 (1981)
7.53 K. Maeda, M. Sato, A. Kubo, S. Takeuchi: J. Appl. Phys. **54**, 161 (1983)
7.54 K. Maeda, S. Takeuchi: J. de Phys. **44**, C4–375 (1983)
7.55 K.H. Küsters, H. Alexander: Physica B **116**, 594 (1983)
7.56 P.J. Dean, W.J. Choyke: Adv. Phys. **26**, 1 (1977)
7.57 L.C. Kimerling: Solid-State Electron **21**, 1391 (1978)
7.58 D. Weeks, J.C. Tully, L.C. Kimerling: Phys. Rev. B **12**, 3286 (1975)
7.59 H. Sumi: Physica B **117 & 118**, 197 (1983); Phys. Rev. B **29**, 4616 (1984)
7.60 K. Maeda, S. Takeuchi: Proc. Yamada Conf. on Dislocations in Solids (University of Tokyo Press, Tokyo 1985) p. 433
7.61 K. Nakagawa, K. Maeda, S. Takeuchi: J. Phys. Soc. Jpn. **49**, 1909 (1980)
7.62 J.S. Nadeau: J. Appl. Phys. **35**, 669 (1964)
7.63 K. Maeda: Oyobutsuri (Appl. Phys.) **46**, 609 (1977) (in Japanese)
7.64 Yu.A. Osip'yan, I.B. Savchenko: JETP Lett. **7**, 100 (1968)
7.65 Yu.A. Osip'yan, V.F. Petrenko: Sov. Phys. – JETP **42**, 695 (1976)
7.66 S. Takeuchi, K. Maeda, K. Nakagawa: *Defects in Semiconductors II*, ed. by S. Mahajan, J.W. Corbett (North-Holland, New York 1983), p. 461
7.67 Yu.A. Osip'yan, I.B. Savchenko: Sov. Phys. – Solid State **14**, 1723 (1973)
7.68 L. Carlsson: J. Appl. Phys. **42**, 676 (1971)
7.69 Yu.A. Osip'yan, V.F. Petrenko: Sov. Phys. – JETP **48**, 147 (1978)
7.70 E.Y. Gutmanus, P. Haasen: Phys. Status Solidi A **63**, 193 (1981)
7.71 Yu.A. Osip'yan, V.F. Petrenko, I.B. Savchenko: JETP Lett. **13**, 442 (1971)
7.72 V.F. Petrenko, R.W. Whitworth: Philos. Mag. A **41**, 681 (1980)
7.73 K. Nakagawa, K. Maeda, S. Takeuchi: J. Phys. Soc. Jpn. **50**, 3040 (1981)
7.74 Yu.A. Osip'yan, V.F. Petrenko, A.V. Zaretskii, R.W. Whitworth: Adv. Phys. **35**, 115 (1986)

Chapter 8

8.1 J. Weertman, J.R. Weertman: In *Physical Metallurgy*, ed. by R.W. Cahn (North-Holland, Amsterdam 1965) p. 793
8.2 J. Weertman: Trans Am. Soc. Met. **61**, 681 (1968)
8.3 M.F. Ashby: Acta Metall. **20**, 887 (1972)
8.4 H.J. Frost, M.F. Ashby: *Rate Processes in Plastic Deformation of Materials*, ed. by J.C. Li, A.K. Mukherjee (American Society of Metals, Metals Park, OH 1975) p. 70
8.5 J. Gittus: *Creep, Viscoelasticity and Creep Fracture in Solids* (Applied Science, Barking, UK 1975) p. 16
8.6 G.B. Gibbs: Mem. Sci. Rev. Met. **62**, 781 (1965)
8.7 F.A. Mohamed, T.G. Langdon: Metall. Trans. **5**, 2339 (1974)
8.8 F.R.N. Nabarro: Rep. of a Conf. of Strength of Solids (Physical Society, London 1948) p. 75
8.9 C. Herring: J. Appl. Phys. **21**, 437 (1950)
8.10 R.L. Coble: J. Appl. Phys. **34**, 1679 (1963)
8.11 A.K. Mukherjee, J.E. Bird, J.E. Dorn: Trans. Am. Soc. Met. **62**, 155 (1969)
8.12 J.G. Harper, J.E. Dorn: Acta Metall. **5**, 654 (1957)
8.13 J.G. Harper, L.A. Shepard, J.E. Dorn: Acta Metall. **6**, 509 (1958)
8.14 T.G. Langdon, F.A. Mohamed: Mater. Sci. Eng. **32**, 103 (1978)

8.15   H. Oikawa: Scr. Metall. **13**, 701 (1979)
8.16   H. Lüthy, R.A. White, O.D. Sherby: Mater. Sci. Eng. **39**, 211 (1979)
8.17   F.A. Mohamed, T.G. Langdon: Acta Metall. **22**, 779 (1974)
8.18   H. Yoshinaga: Keikinzoku (J. Japan Inst. Light Metals) **29**, 528 (1979)
8.19   A. Seeger: Z. Naturforsch. **9a**, 870 (1954)
8.20   J.C. Li: Can. J. Phys. **45**, 493 (1967)
8.21   G.B. Gibbs: Philos. Mag. **13**, 317 (1966)
8.22   D.T. Peterson, R.L. Skaggs: Trans. Met. Soc. AIME **242**, 922 (1968)
8.23   S.R. MacEwen, O.A. Kupcis, B. Ramaswami: Scr. Metall. **3**, 441 (1969)
8.24   A.A. Solomon: Rev. Sci. Instrum. **40**, 1025 (1969)
8.25   C.N. Ahlquist, W.D. Nix: Scr. Metall. **3**, 679 (1969)
8.26   C.N. Ahlquist, W.D. Nix: Acta Metall. **19**, 373 (1971)
8.27   K. Toma, H. Yoshinaga, S. Morozumi: J. Jpn. Inst. Met. **17**, 102 (1976)
8.28   I. Gupta, J.C.M. Li: Metall. Trans. **1**, 2323 (1970)
8.29   I. Gupta, J.C.M. Li: Mater. Sci. Eng. **6**, 20 (1970)
8.30   H. Yoshinaga, S. Matsuo, H. Kurishita: Trans. Jpn. Inst. Met. **26**, 423 (1985)
8.31   J.C.M. Li: In *Dislocation Dynamics*, ed. by A.R. Rosenfield, G.T. Hahn, A.L. Bement, Jr.,
       R.I. Jaffee (McGraw-Hill, New York 1968) p. 87
8.32   H.S. Chen, J.J. Gilman, A.K. Head: J. Appl. Phys. **35**, 2502 (1964)
8.33   H. Yoshinaga: Bull. Jpn. Inst. Met. **16**, 197 (1977)
8.34   K. Abe, H. Yoshinaga, S. Morozumi: Mater. Sci. Eng. **41**, 65 (1979)
8.35   K. Toma, H. Yoshinaga, S. Morozumi: Trans. Jpn. Inst. Met. **17**, 102 (1976)
8.36   H. Oikawa, K. Sugawara: Scr. Metall. **12**, 85 (1978)
8.37   H. Oikawa: Philos. Mag. A **37**, 707 (1978)
8.38   H. Oikawa, M. Nakata, S. Karashima: Mater. Sci. Eng. **60**, 247 (1983)
8.39   H. Yoshinaga, Z. Horita, H. Kurishita: Acta Metall. **29**, 1815 (1981)
8.40   R.W. Bailey: J. Inst. Met. **35**, 27 (1926)
8.41   E. Orowan: Z. Phys. **89**, 614 (1934)
8.42   K. Abe, H. Yoshinaga, S. Morozumi: Trans. Jpn. Inst. Met. **18**, 479 (1977)
8.43   T.H. Alden: Metall. Trans. A **8**, 1675 (1977)

Chapter 9

9.1   H. Yoshinaga, K. Toma, S. Morozumi: J. Jpn. Inst. Met. **39**, 626 (1975)
9.2   R. von Mises: Z. Angew. Math. Mech. **8**, 161 (1928)
9.3   P.B. Hirsch, D.H. Warrington: Philos. Mag. **6**, 735 (1961)
9.4   J. Friedel: *Dislocations* (Pergamon, Oxford 1964)
9.5   H.E. Evans, G. Knowles: Acta Metall. **25**, 963 (1977)
9.6   F. Garofalo: Trans. AIME **227**, 351 (1963); *Fundamentals of Creep and Creep Rupture in
       Metals* (MacMillan, London 1965)
9.7   C.R. Barrett, W.D. Nix: Acta Metall. **13**, 1247 (1965)
9.8   W.D. Nix: Acta Metall. **15**, 1079 (1967)
9.9   W.A. Coghlan, W.D. Nix: Metall. Trans. **1**, 1889 (1970)
9.10   A.H. Cottrell: In *Dislocations and Mechanical Properties of Crystals*, ed. by J.C. Fischer, W.G.
       Johnston, R. Thomson, T. Vreeland, Jr. (Wiley, New York 1957) p. 509
9.11   J. Weertman: Trans. AIME **233**, 2068 (1965)
9.12   J. Weertman: Acta Metall. **15**, 1081 (1967)
9.13   M. Malu, J.K. Tien: Acta Metall. **22**, 145 (1974)
9.14   C.N. Ahlquist, W.D. Nix: Acta Metall. **19**, 373 (1971)
9.15   P.W. Davies, G. Nelmes, K.R. Williams, B. Wilshire: Met. Sci. J. **7**, 87 (1973)
9.16   K. Toma, H. Yoshinaga, S. Morozumi: Trans. Jpn. Inst. Met. **17**, 102 (1976)
9.17   R.W. Bailey: J. Inst. Met. **35**, 27 (1926)
9.18   E. Orowan: Z. Phys. **89**, 614 (1934)

9.19 J. Weertman: J. Appl. Phys. **26**, 1213 (1955)

9.20 J. Weertman: J. Appl. Phys. **28**, 362 (1957)

9.21 J. Weertman: Trans. ASM **61**, 681 (1968)

9.22 L.I. Ivanov, V.A. Yanushkevich: Fiz. Met. Metalloved. **17**, 112 (1964) [Phys. Met. Metallogr. **17**, 102 (1964)]

9.23 W. Blum: Phys. Status Solidi B **45**, 561 (1971)

9.24 J.P. Poirier: *Plasticité à Haute Température des Solides Cristallins* (Editions Eyrolles, Paris 1976)

9.25 K. Maruyama, S. Karashima: Trans Jpn. Inst. Met. **16**, 671 (1975)

9.26 W.J. Evans: Met. Sci. **10**, 170 (1976)

9.27 S.K. Mitra, D.McLean: Proc. R. Soc. London A **295**, 288 (1966)

9.28 D. McLean: Trans. AIME **242**, 1193 (1968)

9.29 H.E. Evans: Philos. Mag. **28**, 227 (1973)

9.30 M.R. Staker, D.L. Holt: Acta Metall. **20**, 569 (1972)

9.31 A. Orlovà, Z. Tobolovà, J. Čadek: Philos. Mag. **26**, 1263 (1972)

9.32 D. Caillard, J.L. Martin: *Creep and Fracture of Engineering Materials and Structures*, ed. by B. Wilshire, D.R.J. Owen (Pineridge, Swansea 1981) p. 17

9.33 D. Caillard, J.L. Martin: Acta Metall. **30**, 791 (1982)

9.34 A.S. Argon, S. Takeuchi: Acta Metall. **29**, 1877 (1981)

9.35 J.D. Parker, B. Wilshire: Philos. Mag. **34**, 485 (1976)

9.36 J.D. Parker, B. Wilshire: Philos. Mag. **41**, 665 (1980)

9.37 D. McLean: Rep. Prog. Phys. **29**, 1 (1966)

9.38 S.K. Mitra, D. McLean: Met. Sci. J. **1**, 192 (1967)

9.39 D. McLean: Trans. AIME **242**, 1193 (1968)

9.40 A. Odèn, E. Lind, R. Lagneborg: *Creep Strength in Steel and High-Temperature Alloys* (The Metals Society, London 1974), p. 60

9.41 T.H. Alden: Philos. Mag. **25**, 785 (1972)

9.42 U.F. Kocks: Philos. Mag. **13**, 541 (1966)

9.43 T.H. Alden: Metall. Trans. **4**, 1047 (1973)

9.44 T.H. Alden: Metall. Trans. A **6**, 1597 (1975)

9.45 T.H. Alden: Metall. Trans. A **8**, 1675, 1857 (1977)

9.46 T. Watanabe, S. Karashima: Met. Sci. J. **4**, 52 (1970)

9.47 C.R. Barrett, C.N. Ahlquist, W.D. Nix: Met. Sci. J. **4**, 41 (1970)

9.48 Z. Horita, H. Yoshinaga: J. Jpn. Inst. Met. **44**, 1273 (1980)

9.49 Y. Ishida, D. McLean: J. Iron Steel Inst. **205**, 88 (1967)

9.50 H. Oikawa, M. Nakata, S. Karashima: Mater. Sci. Eng. **60**, 247 (1983)

9.51 H. Nakashima, M. Watase, H. Yoshinaga: Trans. Jpn. Inst. Met. **27**, 122 (1986)

9.52 S. Sakurai, K. Abe, H. Yoshinaga, S. Morozumi: J. Jpn. Inst. Met. **42**, 432 (1978)

9.53 H. Oikawa, A. Goto, S. Karashima: Trans. Jpn. Inst. Met. **21**, 15 (1980)

9.54 H. Oikawa, M. Ohnuma: Metall. Trans. A **12**, 1699 (1981)

9.55 Y. Adda, J. Philibert: *La Diffusion dans les Solides* (Presses Universitaires des France, Paris 1966) p. 1129

9.56 A.W. Cochardt, G. Schoek, H. Wiedersich: Acta Metall. **3**, 533 (1955)

9.57 G. Schoek, A. Seeger: Acta Metall. **7**, 469 (1959)

9.58 A.H. Cottrell, M.A. Jaswon: Proc. R. Soc. London A **199**, 104 (1949)

9.59 H. Asada, R. Horiuchi, H. Yoshinaga, S. Nakamoto: Trans. Jpn. Inst. Met. **8**, 159 (1967)

9.60 K. Matsuura, T. Nishiyama, S. Koda: J. Jpn. Inst. Met. **31**, 1042 (1967)

9.61 T. Takeyama, H. Takahashi: Trans. Iron Steel Inst. Jpn. **13**, 293 (1973)

9.62 H. Yoshinaga, K. Toma, K. Abe, S. Morozumi: Philos. Mag. **23**, 1387 (1971)

9.63 R. Horiuchi, H. Yoshinaga, S. Hama: Trans. Jpn. Inst. Met. **6**, 123 (1965)

9.64 R. Horiuchi, M. Otsuka: J. Jpn. Inst. Met. **35**, 406 (1971)

9.65 C.M. Sellars, A.G. Quarrel: J. Inst. Metals **90**, 329 (1962)

9.66 O.D. Sherby, P.M. Burke: In *Progress in Materials Science*, Vol. 13, ed. by B. Chalmers, W. Hume-Rothery (Pergamon, Oxford 1966) p. 325

9.67 H. Suzuki: *Ten-i-ron Nyumon* [Introduction to Dislocation Theory] (Agune, Tokyo 1967) in Japanese

9.68 R. Horiuchi, H. Yoshinaga: Trans. Jpn. Inst. Met. **6**, 131 (1965)

9.69 A.H. Cottrell, B.A. Bilby: Proc. Phys. Soc. A **62**, 49 (1949)

9.70 H. Yoshinaga, S. Morozumi: Philos. Mag. **23**, 1351 (1971)

9.71 H. Yoshinaga: Bull. Jpn. Inst. Met. **10**, 519 (1971)

9.72 H. Yoshinaga, S. Morozumi: Philos. Mag. **23**, 1367 (1971)

9.73 M. Sakamoto: Bull. Jpn. Inst. Met. **20**, 912 (1981)

9.74 J.P. Hirth, J. Lothe: *Theory of Dislocations* (McGraw-Hill, New York 1968) p.584

9.75 H. Yoshinaga: *Hagane no Nekkankako no Kinzokugaku* [Metallurgy on Hot Working of Steels] (The Iron and Steel Institute of Japan, Tokyo 1982) p.13, in Japanese

9.76 W.G. Johnston: J. Appl. Phys. **33**, 2716 (1962)

9.77 H. Yoshinaga: *Kouonhenkei to Kouonhakai* [Deformation and Fracture at High Temperatures] (The Iron and Steel Institute of Japan, Tokyo 1981) p.7, in Japanese

9.78 H. Oikawa, N. Kuriyama, D. Mizukoshi, S. Karashima: Mater. Sci. Eng. **29**, 131 (1977)

9.79 P. Yavari, T.G. Langdon: Acta Metall. **30**, 2181 (1982)

9.80 J. Weertman, J.R. Weertman: In *Physical Metallurgy*, ed. by R.W. Cahn (North-Holland, Amsterdam 1965) p.793

9.81 H. Yoshinaga, S. Matsuo, H. Kurishita: Trans. Jpn. Inst. Met. **26**, 423 (1985)

9.82 H. Hayakawa, H. Nakashima, H. Yoshinaga: J. Jpn. Inst. Met. **53**, 1113 (1989)

9.83 G. Saada: Acta Metall. **8**, 841 (1960)

9.84 M. Otsuka, R. Horiuchi: J. Jpn. Inst. Met. **36**, 809 (1972)

9.85 M.J. Mills, J.C. Gibeling, W.D. Nix: Acta Metall. **33**, 1503 (1985)

9.86 S. Takeuchi, A.S. Argon: Acta Metall. **24**, 883 (1976)

9.87 H. Oikawa, J. Kariya, S. Karashima: Met. Sci. **8**, 106 (1974)

9.88 H. Yoshinaga, K. Toma, S. Morozumi: Trans. Jpn. Inst. Met. **17**, 559 (1976)

9.89 E. Peissker, P. Haasen, H. Alexander: J. Appl. Phys. **30**, 129 (1959)

Chapter 10

10.1 E. Orowan: Discussion in the Symp. on Internal Stresses in Metals and Alloys (Institute of Metals, London 1948) p.451

10.2 A.J.E. Foreman, M.J. Makin: Philos. Mag. **14**, 911 (1966)

10.3 U.K. Kocks: Acta Metall. **14**, 1629 (1966)

10.4 L.M. Brown, R.K. Ham: *Strengthening Methods in Crystals*, ed. by A. Kelly, R.B. Nicholson (Elsevier, Amsterdam 1971) p.12

10.5 R. Lagneborg, B. Bergman: Metal Sci. **10**, 20 (1976)

10.6 G.S. Ansell, J. Weertman: Trans. AIME **215**, 838 (1959)

10.7 R. Blickensderfer: Metall. Trans. **5**, 2347 (1974)

10.8 R.S.W. Shewfelt, L.M. Brown: Philos. Mag. **35**, 945 (1977)

10.9 R.S.W. Shewfelt, L.M. Brown: Philos. Mag. **30**, 1135 (1974)

10.10 R. Lagneborg: J. Mater. Sci. **3**, 596 (1968)

10.11 R. Lagneborg: Scr. Metall. **7**, 602 (1973)

10.12 R.W. Lund, W.D. Nix: Acta Metall. **24**, 469 (1976)

10.13 M. Malu, J.K. Tien: Acta Metall. **22**, 145 (1974)

10.14 R.W. Lund, W.D. Nix: Metall. Trans. **6A**, 1329 (1975)

10.15 E. Arzt, M.F. Ashby: Scr. Metall. **16**, 1285 (1982)

10.16 M. McLean: Acta Metall. **33**, 545 (1985)

10.17 H. Nakashima, K. Iwasaki, S. Goto, H. Yoshinaga: J. Jpn. Inst. Met. **52**, 180 (1988)

10.18 D.J. Srolovitz, M.J. Luton, R. Petkovic-Luton, D.M. Barnett, W.D. Nix: Acta Metall. **32**, 1079 (1984)

10.19 R.W. Weeks, S.R. Pati, M.F. Ashby, P. Barrand: Acta Metall. **17**, 1403 (1969)

10.20 R. Horiuchi: Symposium on Microscopic Approach to the Mechanical Properties of Materials, held by Japan Inst. Metals (1986)

10.21 J.H. Schröder, E. Arzt: Scr. Metall. **19**, 1129 (1985)

10.22 E. Arzt, D.S. Wilkinson: Acta Metall. **34**, 1893 (1986)

10.23 S. Goto, S. Yamashita, T. Mimura, H. Yoshinaga: Trans. Jpn. Inst. Met. **27**, 512 (1986)

10.24 H. Kurishita, H. Yoshinaga, F. Takao, S. Goto: J. Jpn. Inst. Met. **44**, 395 (1980)

10.25 R.D. Schmidt-Whitley: Z. Metallkd. **64**, 552 (1973)

10.26 W.S. Williams, R.D. Schaal: J. Appl. Phys. **33**, 955 (1962)

10.27 K. Hara, H. Yoshinaga, S. Morozumi: J. Jpn. Inst. Met. **42**, 1039 (1978)

10.28 H.E. Cline, D.F. Stein: Trans. AIME **245**, 841 (1969)

Subject Index